Dr Dave Elliott is Professor of Technology Policy in the Faculty of Technology at the Open University and Director of the OU Energy and Environment Research Unit. He trained initially as a physicist and worked for the UK Atomic Energy Authority at Harwell and the Central Electricity Generating Board in Bristol.

At the Open University he has been looking at energy policy issues and in particular at renewable energy policy. He is co-ordinator of the Network for Alternative Technology and Technology Assessment (NATTA) and editor of its journal, *Renew*. In addition to writing many texts for the OU, he has published over 50 academic papers and reports. He has also authored, co-authored or edited eight books. Routledge have just published an enlarged and updated version of his best-selling book *Energy, Society and Environment*.

He was the recipient of the 1996 Schumacher Award.

Schumacher Briefing No. 10

A SOLAR WORLD

Climate Change and the Green Energy Revolution

David Elliott

published by Green Books
for The Schumacher Society

First published in 2003
by Green Books Ltd
Foxhole, Dartington, Totnes,
Devon TQ9 6EB
www.greenbooks.co.uk
greenbooks@gn.apc.org

for The Schumacher Society
The CREATE Centre, Smeaton Road,
Bristol BS1 6XN
www.schumacher.org.uk
admin@schumacher.org.uk

Cover design by Rick Lawrence

Printed by J. W. Arrowsmith Ltd, Bristol, UK

A catalogue record for this publication
is available from the British Library

ISBN 1 903998 31 X

Contents

Foreword

Whether we run out of fossil fuels sooner rather than later, the very fact of burning these fuels in vast quantities is putting the future of people and planet into question. Climate change is upon us, and three responses are required: efficient energy use, the rapid phasing in of renewable energies, and structural changes in the way we live and run our economy. Dave Elliott's penetrating analysis of the options open to us arises out of years of leading-edge work in the field of renewable energy. As co-director of EERU, the Energy and Environment Research Unit at the Open University, as author of many books and as editor of *Renew*, the newsletter on technology for a sustainable future, he has a track record second to none. He received a Schumacher Award for his work in 1996.

A Solar World is a *tour de force* survey of the rapidly growing field of green energy. In 96 pages of text, it covers a tremendous range of issues: after setting the scene in the context of climate change, and a strong rebuttal of arguments promoting nuclear energy, Dave launches into a powerful argument for bringing renewable energy centre stage. He discusses the critical issue of security of energy supplies in a world where renewable energy will be a major contributor. He discusses the importance of systems integration and the potential for fuel cell technology in this context. He looks at the cost implications of the green energy revolution. He raises the question of ownership, particularly of wind farms, and the issue of the social control of future energy systems. What can we do to assure that people want to live with renewable energy, and particularly wind power, rather than protest against it?

Dave then raises the thorny issue of whether renewables can provide sufficient energy to run our transport systems. Are major changes required in the way we operate transport in the context of a globalising economy, or is localisation a necessary response to our profligate use of fossil fuels? Can hydrogen be used in significant quantities as the main transport fuel in a sustainable future?

A particularly interesting aspect of the Briefing is that it looks at renewable energy in the wider context of a move towards a sustainable society. But these are not utopian dreams. Much of the final section of the Briefing is dedicated to a detailed discussion of the policy drivers now being used in some countries to make the rapid introduction of green energy a reality. Will the UK catch up with those countries in Europe that have taken the lead, or will our obsession with energy costs keep us signed up to polluting, climate change-inducing fossil fuels? It is clear that supporting renewables is of critical importance in a transition period, but it does not need to cost the exchequer vast amounts of money.

Dave Elliott has written an important text for all of us concerned about how long the 'age of fire' can continue. It is for concerned individuals as well as for decision-makers who are still sitting on the fence. I trust that, as with previous Briefings, Dave's short book will make an important contribution on how to create a sustainable, human-scale future.

Herbert Girardet,
Commissioning Editor,
Schumacher Briefings

Chapter 1

Introduction

Sustainable energy

Renewable energy has moved from being seen as a marginal, quirky, concern to being centre stage. For example, Denmark now generates around 18% of its electricity from wind turbines and the European Union as a whole aims to generate 22% of its electricity from renewable energy sources by 2010.[1] And around the world, the case for a switch to using the winds, waves, tides, direct solar and biological sources of energy, gets stronger every day—as the impending problems of climate change, due to the use of fossil fuels, become clearer. Not only do advocates of sustainable energy claim that the development of renewable energy will help us avoid environmental and resource scarcity problems, they also sometimes claim that this switch could be relatively easy. Indeed, some suggest that it will be a case of 'win-win'. Far from imposing costs, the change over to sustainable approaches to energy, and indeed to all parts of the economic system, will open up opportunities for new jobs and economic expansion.

For example, speaking at a CBI-Greenpeace Business Conference in October 2000, Stephen Byers, then UK Secretary of State for Trade and Industry, suggested that the UK could and should embark on what he called a 'Green Industrial Revolution'. He described a 'win-win' vision of prosperity through environmental modernisation. 'We can not continue to pollute the environment and consume resources in the way we have over the last two hundred years. We all know that is unsustainable. We therefore need demanding, long term objectives and goals to improve our productive use of resources, and to cut waste and pollution. That is not a threat to business, to growth, to prosperity. It is the key to our future prosperity.'

Prime Minister Tony Blair underlined this commitment in March 2001, arguing that the UK's investment in renewable energy technology was 'a major down-payment on our future, and will help open up huge commercial opportunities for Britain'.

In this Briefing I want to look at whether we really can hope to experience 'all gain and no pain', as some suggest, or whether coming to terms with the industrial and consumer system we have created will require, at the very least, some difficult trade-offs between social and environmental objectives.

Given the space available, rather than taking on the whole industrial system, I will focus just on the energy sector and energy use. That is not an inappropriate restriction, since the current pattern of energy use underpins most of our environmental problems, most notably climate change.

Improving the efficiency with which we use energy is one of the key options for limiting the impact of climate change. The potential for energy saving is huge, and I will be looking at some examples. There is certainly a lot that can be done using fairly simple and cheap technical adjustments to the way we use energy. However, even if we avoid wasting so much energy, and begin to cut back on the seemingly inexorable rise in demand for energy, there will still be a need for cleaner energy supplies, and my main emphasis will be on renewable energy sources—the vital new ingredient to a sustainable future.

Box 1: Energy Units: a few terms and statistics you may find useful

kW = kilowatt (1000 watts) A typical one bar electric fire has a rated power of 1kW. Run for one hour it would consume 1 kilowatt hour (kWh) of energy

MW =megawatt (1000kW) Modern wind turbines are rated at 1-2 MW

GW = gigawatt (1000MW) Large coal or nuclear plants are around 1GW

The UK consumes about 350,000,000,000kWh of electricity per annum, or 350 Terawatts ('TW'). That's about a third of the UK's total energy consumption—the rest is used in the form of heat (mostly from gas) and transport (mostly from petrol).

The UK's overall energy consumption has been increasing at about 1-2% per annum, with the transport sector taking the lead.

Why we need to change

The climate change problem can be stated quite simply. At one time, one of the key energy issues was how long fossil fuel reserves would last. That is still a major issue, but it has been overtaken by the realisation that we may not be able to burn off whatever reserves are left without damaging the climate system. It took hundreds of millions of years for the carbon dioxide that was once in the primeval atmosphere to be converted into fossil materials underground. But we have burnt off a good part of it over a hundred years or so of ever-increasing economic activity—mostly in power stations and vehicles. Carbon dioxide levels in the atmosphere are now higher than they have ever been in recorded history—even history recorded in the ice cores going back hundreds of thousands of years. The parallel rise in average planetary temperature seems to be the inevitable result, and the impact of this continuing process on life on earth could be huge.[2]

It's not just a matter of a global warming. A 5 or 6°C rise over the next century would begin to melt the icecaps and that, coupled with the thermal expansion of the seas, would probably lead to sea level rise of half a meter or more, inundating many key food-growing and urban areas of the word. The weather system would become more erratic, with an increase in major storms and floods, and, at other times, droughts. Some more pessimistic projections suggest that global warming could kill off the rainforests, so that their stored carbon dioxide would trigger a runaway greenhouse effect, with methane from the seas accelerating it further. As a result there could be temperature rises of up to 20°C.

None of this is necessary. We can use the climate and weather system to generate energy without producing the emissions that damage the climate system. The natural energy flows in the winds, waves, river currents and tides, the solar-driven biological energy in biomass and the use of direct solar heat and light can provide for all our needs, if we can devise the necessary energy conversion technologies and use energy efficiently. That is the vision offered by advocates of the sustainable energy path.

In general terms there is no doubt that it would be possible to rely just on renewable sources, without using any coal, oil, gas or nuclear fuels. The sun delivers more energy per square metre of the planet than we could ever need—the resultant natural energy flows represent a very large, non-finite, resource base, with the solar radiation input alone amounting to around 90,000 Terawatts (TW). For comparison, our total global energy consumption, measured on a continuous equivalent power basis, is only around 13 TW.[3] Of course, not all of the 90,000 TW solar input can be successfully captured and used. Most of the flows are diffuse, some are intermittent, and the efficiency of conversion technologies has to be taken into account, as does the location of the source. But even so there is plenty there, if we can devise the technologies.

Fortunately we have been doing that—too slowly maybe, but steadily. There is a long way to go, given that at present fossil fuels supply around 76% of the worlds energy (see Box 2). However, we can now just about see how we could, in principle, move to a sustainable energy system based on the use of renewable energy, coupled with careful attention to the efficient use of energy. Technically it is possible. The big issue is whether we can and will do it, and do it in time to avoid a climate change disaster. An ancillary issue is whether a commitment to renewables and energy efficiency can allow us to avoid the risks of using nuclear power—which is offering itself as another non-fossil option. Let's deal with that first.

The end of nuclear power

It is true that nuclear plants do not directly generate any carbon dioxide gas. But they have a range of other environmental drawbacks. Most environmentalists see the nuclear option as a non-starter, given that it has proved to be expensive and potentially dangerous, and as yet there is no agreed way of storing very long-lived radioactive wastes that are produced. It also opens up a range of weapons proliferation and security risks. I am not going to rehearse all the reasons why nuclear power cannot be relied on to help deal with climate change. I have laid out my views extensively elsewhere.[4] Suffice it to say that, quite apart from the excessive costs, safety and security risks, it seems foolish to try to solve one major environmental problem, climate change, by introducing another, radioactive pollution.

After years of massive funding support, nuclear power only

***Box 2:* Energy around the World**

Global primary energy consumption in 2000, by source:

Hydro	2.3%
Nuclear	6.6%
Oil	34.6%
Biomass	11.3%
Gas	21.4.8%
Coal	21.6%
New renewables	2.1%

See the World Energy Councils web site for regular updates: http://www.worldenergy.org.

Obviously the pattern of energy use for each individual country differs. For example, Norway and Brazil obtain more than half their electricity from hydro, France more than 70% from nuclear power, whereas in some developing countries there is very little use of electricity from any source. The UK gets around 25% of its electricity from nuclear power plants, nearly 30% from coal-fired plants and over 40% from gas-fired plants. Hydro and new renewables produce around 3% of the UK's electricity.

Electricity is not the only type of energy used. For example, in the UK it represents only about 20% of total energy use. In addition to being used to generate electricity, gas is used for heating (representing about 30% of total UK energy used), and oil is used for transport fuels (about 40% of total energy). Coal and other types of solid and liquid fuel, for heating and industrial use, make up the rest.

In overall terms, the industrialised countries, with about 20% of the world population, use about 60% of the world's energy. That translates into an even larger imbalance in per capita terms—the richest billion people use 5 times more energy than the poorest 2 billion. However, some parts of the developing world are beginning to catch up in terms of total national energy consumed per annum. Certainly, in terms of overall energy use globally, the general trend is upward, by around 1-2% extra each year.

provides around 6.6% of the world's energy, and most European countries have no plans for new nuclear plants and are phasing out their existing plants. In its 2003 White Paper on Energy, the UK government concluded that 'its current economics make it an unattractive option and there are also important issues of nuclear waste to be resolved.' It therefore did not contain proposals for building new nuclear plants and focused instead on renewable energy and energy efficiency.[5]

Of course the ex-Soviet states still have nuclear programmes, as do Japan, India and China; and they, and some other Asian countries, would like to expand this option. However, overall the expansion of nuclear power globally seems to be stalled. In the USA, which has not ordered a new nuclear plant since the Three Mile Island partial core meltdown accident in 1979, many of the existing plants are now reaching the end of their useful life and are not being replaced, although, of late, President Bush seems to be trying to breathe life back into the programme.

It is sometimes argued that we can't abandon nuclear power without major problems of energy security. It will be harder for some countries than others, with France, which currently obtains around 75% of its electricity from nuclear plants, probably being the worst case. However most of the French plants are now old and the 'red-green' coalition government seems unlikely to replace them. Indeed it has launched a major renewable energy programme. Belgium, which currently obtains 57% of its electricity from nuclear plants, now has a programme of phasing them out by 2025, while Germany plans to phase out its 19 nuclear plants by around 2030, and has embarked on a major renewable energy programme which, it hopes, will supply around 50% of its electricity by 2050.

These are relatively leisurely phase-out and replacement programmes, and most environmentalists would like to see them accelerated. The UK currently obtains around 25% of its electricity from nuclear plants, most of them being operated by British Energy. Most are scheduled for closure over the next two decades, as they reach the end of their operational life: some have already reached that point. However a faster phase-out is possible. A study by the ILEX consultancy, produced for Greenpeace in 2003, found that, given the UK's large excess generating capacity, all British Energy's nuclear plants could be closed by the winter of 2005/6 without undermining security of supply. The UK would still have an overall 20% generation margin over maximum winter demand. That is hardly surprising—the UK has nearly 80 gigawatts of generation capacity installed, whereas meeting the maximum winter demand in 2002 only required around 59GW. Indeed, ILEX concluded, the UK could close all its nuclear plants, including BNFL's old MAGNOX reactors, by the winter of 2006/7 and still retain a 20% plant margin.[6]

What about the longer term? With nuclear removed, the way ahead for a rapid expansion of renewables would be open, so that, with energy efficiency continuing to make significant contributions, emissions would continue to fall. The removal of nuclear power would of course mean the loss of some carbon-free generation, but the expansion of renewables and energy efficiency should more than compensate for this. It interesting to note that, although 32GW of nuclear generation capacity has so far been phased out around the world, coincidentally, exactly 32GW of wind generation has been installed. Of course, the load factors for wind and nuclear plants differ by around a factor of two, so this is not a full replacement, but given continued expansion of wind, wave, tidal, solar and other renewables, and the development of more efficient ways of using fossil fuel, it should be possible to replace nuclear and reduce emissions. A report produced in 2003 by Friends of the Earth, 'Tackling Climate Change without Nuclear Power,' found that by 2020, a sustainable energy policy, based mainly on renewables and energy efficiency, with nuclear power being phased out, could reduce overall emissions by 45%.[7] That would put the UK well on the way to the 60% reduction by 2050 called for by the Royal Commission on Environmental Pollution as a response to climate change.[8]

A low carbon non-nuclear future for the UK, based increasingly on renewables, is not an eco-enthusiasts' fantasy. It a practical proposition. That was one of the conclusions of the Royal Commission on Environmental Pollution. In his detailed study of 'The UK's Transition to a Low Carbon Economy', Paul Ekins came to similar conclusions.[9] So did the government's Performance and Innovation Unit in its 'Energy Review'.[10] What we now need is the political will and financial commitment to make this approach a reality, rather than trying to keep the nuclear dream alive. The first half of that seems to be apparent in the government's White Paper on Energy published in February 2003—it backed the idea of cutting emissions by 60% by 2050. However, so far, the financial resources have been less forthcoming—despite all its green rhetoric, the White Paper only allocated an extra £60m to renewables.[11]

Unfortunately, a lot of resources are needed to deal with the left-overs of the nuclear programme—for example in 2002 £650m was provided to bail out British Energy. And whatever we now do, there will be around 500,000 tones of radioactive waste of various types to deal with somehow—including the wastes that will be generated by decommissioning the UK's existing plants.[12] Adding to that by investing yet more money in new nuclear plants makes no sense. Instead it would seem sensible to let renewable energy technology have a chance to show what it can do.

Of course, some people are hopeful that nuclear fusion will emerge as a viable option, and solve all our energy problems. Unfortunately, even optimistically, and despite having cost over £20bn so far, the fusion option is decades away and, as yet, the economics are very unclear. Moreover, although there would not be any conventional radioactive waste products, the high neutron radiation in the fusion reactor core would induce radioactivity in the surrounding materials and containment system, which would have to be periodically stripped and stored. So there would still be a waste problem. And working with radioactive plasmas at 200 million degrees centigrade is likely to present some serious safety and security hazards, not least the risks of the release of radioactive tritium gas. All in all, rather than trying to create little suns on earth in the form of fusion reactors, it would be more sensible to make use of the fusion reactor we already have (the sun), which provides the energy for a wide range of continuously renewed energy flows on the earth— including wind, waves and the energy in rivers flows. These renewable sources are ready for us to use now, whereas viable fusion reactors are still something of a science fiction dream and are unlikely to be available in time to help us deal with the urgent problem of climate change.

What can renewables offer?

There are a lot of renewable energy sources. What they share in common is that, as well as having generally low environmental impacts, they are naturally renewed—they will not run out and can continue to sustain us indefinitely. However they all have different characteristics in terms of the scale and location of the technologies needed to use them. Some renewable energy sources, like solar heat, can be tapped at the small scale, for example via solar collectors on individual rooftops. Others are larger scale and will be located in remoter areas, like wave energy devices located out to sea, sending power by marine cable back to land.

A quick snapshot of how and where renewables might be used may help set the scene for the discussion in this Briefing. What follows is inevitably speculative and represents only one set of possibilities. One of the attractions of renewables is that there are lots of them and lots of different ways in which they can be used. What follows is based mainly on examples from industrialised countries like the UK, in part because that is where I have most experience. Later on I will be looking at some of the implications for the rest of the world. But you could argue that countries like the UK, which pioneered the existing energy systems with all their problems, ought to be responsible for pioneering and promoting their replacements.

Let's look at rural areas first. Most renewable energy flows are diffuse, so you need space to collect the energy. For some you need quite a lot of space—energy crops for example. Fast-growing willow coppices can be harvested after two or three years growth to provide wood chips to fuel a power station to generate electricity. But you will need quite a lot of land to provide sufficient fuel for a reasonably sized plant—something like 100-200 hectares for each megawatt. You would need even more land if you want to grow fuel for vehicles, since the energy conversion ratios (energy out per energy input) for liquids, like green diesel from rape seeds, are lower than for and solid energy crops like willow. Wind farms also occupy relatively large areas, although the turbine bases themselves only occupy a few percent of the farm area, and the rest of the land can be used for conventional farming. And overall, wind farms require about 20 times less land per unit of energy produced than energy crop-based systems.[13]

As you can see though, all these are basically rural technologies. Some energy crops might be grown in urban areas and some wind projects may find locations in cities, but in the main, the bulk of the power generated from these sources is going to come from rural areas. The same is obviously the case for large-scale hydro and large tidal barrage projects. Although some micro hydro, and maybe even micro tidal projects, could be located in suitable urban areas, most are likely to be in remote areas. The offshore options—like offshore wind, wave and tidal current devices—will also mostly tend to be located off the coast in remote parts of the country.

Taken together these various rural and offshore resources could provide much more energy than is needed by rural populations. Which is just as well, because urban areas will probably not be able to generate sufficient for their needs. I say 'probably', because that rather depends on whether those needs remain unchanged.

At present around 80% of the UK population lives in urban areas of various types. Like their fellows in other parts of the country, many urban residents live in houses which are relatively inefficient from an energy point of view. Most of the energy fed in is wasted, due to poor building design and the use of inefficient energy-using devices. That can be changed. In principle, for example, given the necessary building regulations and planning controls, there is no reason why the energy used in houses for heating could not be cut very dramatically. There are super-insulated buildings which use one-tenth of the energy used by many current UK dwellings. Given proper performance standards and regulatory controls on the equipment supplied by manufacturers, savings can also be made by the adoption of more energy-efficient devices in the home, for lighting, heating, cooking and so on. Similar savings can be made in commercial and retail premises. So, overall, urban energy demand could be reduced, making it easier to meet from renewables. **See Box 3.**

That could be somewhat more credible when you realise that, although urban areas may not have much free land, they do have large amounts of roof space, much of it orientated conveniently to the south, so that it can be used to house solar collectors for heating and photovoltaic solar modules (i.e. arrays of solar cells) for

Box 3: Green cities?

Herbie Girardet's Schumacher Briefing No. 2 *Creating Sustainable Cities* provides a useful overview of some of the ways in which cities can become more sustainable, with examples from around the world, including the celebrated case of Curitiba in Brazil which has developed an excellent integrated transport system. Transport is obviously one of the big environmental issues for cities, not least since it involves energy use, but so too is direct energy use for heating and lighting.

Some progress is being made. For example, studies of Leicester carried out by the Open University Energy and Environment Research Unit have suggested that energy demand could be cut by at least 60%, and much of this reduced energy requirement could then be met by CHP plants and from locally available renewable sources, with renewables supplying perhaps around 25% of the city's power. As a result, by 2020, Leicester's CO_2 emissions could be cut by nearly 80% on 1990 levels. One response to studies like this was a commitment, in Leicester's Local Agenda 21 Strategy, to try to obtain 15% of total energy production in the county from renewable energy sources by the year 2020.

Many other urban areas have responded similarly. The area with the largest problem is of course London, but even there progress is being made. The Mayor of London's Energy Strategy 'Green Light for Clean Energy' published in January 2003 as a consultative document, suggests that 14% of London's electricity could and should come from renewable energy sources by 2010, although it accepts that some of this would have to be imported. Even so it paints an interesting vision of the future: 'At least 10,000 domestic photovoltaic schemes should be installed in London, as well as 100 photovoltaic applications on commercial and public buildings and 6 large wind turbines, 500 small wind generators associated with public or private sector buildings, 25,000 domestic solar water heating schemes, 2,200 solar water heating schemes associated with swimming pools as well as more anaerobic digestion plants with energy recovery and biomass-fuelled combined heat and power plants'.

Solar energy clearly makes a lot of sense for cities, given the large roof areas available, and as the examples given above illustrate, this is not just limited to sunny southern locations. In fact, the annual total solar irradiation levels in equatorial areas (2000 kilowatt-hours per square metre of horizontal surface) are only twice those in northern latitudes like the UK—and, leaving air conditioning aside, in the latter the value of the heat will be much higher. Moreover, PV solar arrays do not need heat, just light, so they will work even without strong direct sunlight.

electricity production. It has been calculated that most cities have enough suitable roof space to meet a significant proportion of their energy needs from solar heat and solar electricity systems. Obviously the availability of this energy source is cyclic, but solar energy is well matched to daytime energy use in offices and retail outlets, and is also well suited to meeting cooling and air conditioning loads in the summer.[14]

However, solar power clearly won't provide a continuous direct input around the clock or around the year; other sources will be needed at times to top up. Fortunately there are some other urban options. What cities also have, in addition to rooftops, is large amounts of domestic and municipal waste material, much of which can be used as an energy source—although only after all the useful materials have been recycled. Most environmentalists see waste combustion even for energy recovery as a bad idea, but there are many other options, including the new technologies of pyrolysis and gasification, which should avoid most of the environmental problems associated with conventional 'mass burn' waste incineration.

Some of these new waste conversion technologies can be used to produce methane gas or even hydrogen gas, rather than electricity. These gases have the attraction that they can be stored for later use, so that they can partly balance out the variations in solar availability. We already make use of methane from some urban sewage treatment plants to generate electricity. It could be that we will also begin to make use of other locally produced sources of methane, including wastes, via anaerobic 'biogas' digesters. Thus we may gradually be able to replace the gas we currently get from the diminishing reserves in the North Sea with a renewable source derived from urban sources. In addition, as noted earlier, there may also be opportunities for wind power in urban areas—some new wind turbine designs are emerging which have wind rotors mounted horizontally along the roof ridges of buildings or vertically integrated into the building design. In addition, in some locations, geothermal energy from deep underground can be used. Paris already has over 50 district heating networks running off geothermal heat sources, while Southampton's district heating network also makes use of geothermal heat.

Even so, it does seem likely that urban areas will have to import some power from rural areas. That is of course what happens already. Most coal, nuclear and gas-fired power stations are in open country. However a switch over to the use of renewable energy sources could have a significant effect on rural landscapes in some areas, with wind farms being the most obvious example. Cities, even in the future, will inevitably rely on rural areas for most of their resources—most obviously food and water.

There has, in the past, been resentment from rural people about the flooding of rural areas to create reservoirs to supply water to urban areas. Currently this is perhaps most strongly shown in the anger of rural people in the construction of the huge Tehri dam in Himal Pradesh, India, which is being built primarily for Delhi's benefit. In the UK some of the opposition to wind farms reflects similar resentment. Why, some may say, should rural areas be blighted for the benefit mainly of people in the cities? [15]

It could be then that there will be some interesting new twists to the rural versus urban debate, as and when renewables become widely used. I will be looking at the wind farm debate in detail later. It is one example of the problems that I feel we will have to address as renewables are deployed on a significant scale.

Putting it all together?

So far I have just talked in general terms about supplying rural and urban areas with energy generated from renewable sources. I have focussed mainly on electricity production, although clearly some renewable systems can also provide heat, most obviously solar collectors. These can offer a lot of potential for fuel saving, not least in the domestic sector. Moreover, this is one area where it is relatively easy to have a direct impact yourself—and even to build and install the equipment yourself: see Box 4.

Supplying electricity on a d-i-y basis is harder. While, as we will see, there are opportunities for domestic-scale generation with, for example, rooftop PV solar, and local level generation could expand, for the moment most electricity will be obtained via national power grid systems. However, it seems clear that, overall, renewable sources of various types and scales could ultimately supply most of the electricity we need. On-land wind could ultimately meet perhaps 10% of our electricity needs, while the offshore wind resource could supply at least 20% of UK electricity (possibly much more), and wave energy a similar percentage. Tidal current systems might supply 20%, as could tidal barrages, should we wish to follow that option up. On top of that we have solar PV, geothermal, and small hydro, along with the various biomass options. So there is no shortage of potential. But how would the whole system work? Even assuming a sensible approach to the efficient use of energy, is it really possible to meet all our energy needs from these sources? Wouldn't there be a need for more radical changes in how we use energy—and how we live?

It is certainly true that the way in which we generate and use energy would have to change. But, as it happens, changes are already underway. After the second world war power stations in the UK got larger and larger, with giant 1000 megawatt coal and nuclear plants becoming the norm, feeding the national power grid. There were definitely economies of scale in going bigger and bigger, especially since the power stations were all run by one large public company, the Central Electricity Generating Board. But now, following the break up of the old CEGB into smaller private companies, a new pattern of generation has emerged. Privatisation in 1990 coincided with the arrival of a new technology—small gas-fired turbines. These were cheap and quick to build and they used gas, which was much cheaper than coal. The 'dash for gas' has seen small- to medium-sized Combined Cycle Gas Turbines, which are much more efficient than conventional coal-fired plants, become the dominant energy technology.

So we have begun to switch over to smaller plants, in the range of 20-100MW, a trend which has been continued with the arrival of wind farms, most of which are in the range 20-50MW. Increasingly it is being argued that, in some cases, economies of scale exist for small energy generators. They can meet local energy needs from local resources, and embedding power plants in local area grids avoids the large energy losses (of up to 10% of the power) associated with sending power from large plants long distances on the national grid, as at present.

As I have indicated, we have at last begun to pay more attention to the efficiency with which energy is generated. Conventional

Box 4: Doing it yourself: domestic solar hot water

A solar heat collector installed on your rooftop can typically cut your home heating bills in half. However, gas heating is so cheap that the solar alternative can look expensive, even though the reason for this is that you do not have to pay environmental costs of using gas—at least not directly, yet, in cash. But the costs of going solar can be reduced by co-operative efforts—for example by participating in a group which can bulk buy the necessary materials. Moreover, not everyone may feel well enough equipped technically to build and install a solar collector themselves, in which case a group may be able to provide the necessary skills.

The grass roots self-build solar collector initiative in Austria is a celebrated example. It started in the 1980s, and led to the development by local activists, motivated by the need to get access to cheap and efficient solar technology, of a novel solar heat collector design. The development of the design was incremental, with feedback from initial users being incorporated as it went along, and helped create a version which was taken up by 100,000 users, making Austria one of the leaders in the use of solar power. By the end of 1999 Austria had 2 million square metres of solar collectors installed, of which about 400,000 m^2 were self-build systems.

In a similar way, a dozen or so d-i-y 'Solar Clubs' emerged in the UK during the late 1990s, run by people who wanted to get access to solar energy devices to install on their homes, but could not find what they need, at least not cheaply, on the market. By forming a 'self help' Solar Club, a group of individuals could share skills, labour and purchasing power, thus cutting costs and making d-i-y installation easier. The Club could also be able to get access to information and, possibly, financial support from aid agencies. Indeed several community organisations emerged to help support the spread of such clubs. Once established, the clubs pass on their experiences and expertise to new members. It's self-help sustainability.

Some UK contacts: Centre for Sustainable Energy, Bristol
http://www.cse.org.uk Leicester Environ, http://www.environ.org.uk
SEA London: http://www.sustainable-energy.org.uk/
and http://www.solarforlondon.org/

For the Austrian experience see:
http://www.aee.at/verz/english/self01.html

power plants were only at best 35% efficient: most of the energy in the fuel ended up as waste heat, pumped out into the environment from the familiar huge cooling towers. The new breed of Combined Cycle Gas Turbines are much better, with efficiencies of 55% or more, since they use some of the waste heat to generate more electricity. But we can do even better than that, if we use the waste heat directly, for example to feed into district heating networks. Operating power plants in this way, as Combined Heat and Power (or 'CHP') suppliers, can more than double their overall energy conversion efficiency. CHP plants can be at various scales, right down to small gas-fired domestic units—known as 'micro CHP'—providing both electricity and heat in and for individual buildings.

As we get more serious about the efficient use of energy, we will see more and more systems in which otherwise wasted energy is captured, and many of these systems will be viable on a small scale. We will also be making increasing use of the small 'fuel cell' power units, whose use was pioneered in spacecraft, to provide both heat and power for homes, offices and other buildings.

Given this trend to what has been called 'micropower', it could be that we will move to an energy system in which a lot of the energy needed is generated locally, with any excess being exported to other areas via the national grid, which will also allow 'top up' power to be imported when there is a local shortfall. There would still be a need for a national grid system and for some large centralised power stations to balance out local variations in power availability and demand. Some large wave and tidal projects might but be used in this way. But overall the system would be much more decentralised than at present, with smaller power stations dispersed around the country, right down to the individual house level.[16]

Indeed, it could be that as photovoltaic ('PV') solar modules become cheaper and more widespread, and as other smaller-scale energy systems emerge, using locally produced renewable sources like methane or hydrogen, many buildings will generate most of their energy needs and export excess power—every house a power station! In parallel, in rural areas we might expect local wind farms and micro hydro projects to supply some power to some villages, as well as to the national grid, while biomass projects on the outskirts of towns might also supply heat and power, both locally and nationally.

This new more decentralised pattern of energy generation is already beginning to happen—in part because some people want to generate at least some of their own power so as to be less reliant on grid supplies. This became particularly apparent in California in 2000-2001, when, following the deregulation of the energy market, there were regular blackouts and price hikes. Consumers could no longer trust the grid supplies, so some have turned to PV solar.

The UK has been relatively slow to deploy PV, but, as elsewhere, there are now houses which have PV solar roof arrays which provide power for electrical devices. Any excess is exported to the grid and any shortfall is met by imports from the grid under 'net metering' arrangements—the consumer only pays for the net flow in or out, and in some cases that means they actually make a profit. Of course, for the moment, installing a PV roof is very expensive, but it is interesting to have a roof that pays its way, and one home owner I know of also runs an electric car off the system.[17]

At present there is around 2GW (peak) of PV capacity in place around the world ('peak' refers to the notional full power rating), and there are major programmes of expansion underway, including Germany's 100,000 solar roofs programme. By 2010, as prices fall, Japan expects to have around 5GW (peak) of PV installed.

Certainly, technological developments are leading to quite rapid improvements in efficiency and reductions in price—there are PV devices on sales with efficiencies of nearly 20% and which cost around $5 per (peak) watt—around a tenfold reduction in ten years. If this progress in maintained, PV devices could break through the crucial $1/watt barrier, making them competitive with most other energy sources. When and if that happens, PV could make a major contribution. For example, Greenpeace has estimated that, by 2020, there could be 200GW of PV installed around the world—60% of it in developing countries, much of that necessarily being off-grid.[18]

Security of energy supply

In countries or areas where there are no significant grid links available, independent 'stand alone' generation is the only option, so it is argued that micro power system like PV could be very useful—avoiding the use of imported diesel or propane. As was noted above, the same technology may also be relevant at the domestic level in industrial countries, but, in that case, it would be linked in to the grid system, so as to allow for two-way power trading. It may thus be possible for individuals to run their houses using a micro-power system like PV solar, backed up by the grid, and for the grid to also be fed from larger renewable sources. But would a system like this, based on renewable sources, be as robust and reliable in energy supply terms as the system which we currently enjoy? Surely relying on intermittent sources like solar and wind energy will present major problems if used on a wide scale?

The first thing to note is that not all renewable sources are intermittent. The use of biofuels, hydroelectric power and geothermal energy can provide 'firm' i.e. continuous power, and tidal energy is very predictable. However, the availability of solar, wind, and wave energy is not so predictable.

Fortunately some of the basic annual weather cycles link up well with human energy requirements. For example, as already noted, in many locations, peak daytime air conditioning loads in summer link well with peak power availability from solar photovoltaic systems.

At the other extreme, in much of the world it is windy during the winter so wind and wave power is at a peak when electricity is most needed for heating. Indeed there is a further correlation. Wind produces a chilling effect on buildings and conventional power systems have to take this into account by having extra generating capacity ready for cold windy days. In the UK this can be up to 1GW. However if we have 1GW of wind capacity on the system, its output will be positively correlated to some degree to the increased demand from the wind chill effect.

The shorter term variations in renewable energy inputs are more of a problem, since there can also be negative correlations between availability of renewable energy like wind and energy demand. For example, the wind does not blow all the time, so some wind turbines will be unproductive at any particular time, and this may coincide with cold weather.[19]

This sounds as if it could be a major problem. However, in practice, up to a certain level this intermittency need not be too much of an issue. If the electricity from these devices is fed into an integrated national power grid network, this can 'even out' and 'buffer' the local variations in energy inputs. Individual wind turbines may be becalmed occasionally, but in a country like the UK it is usually windy somewhere, so other wind plants will be delivering power to the grid, a situation that will be further improved as wind farms are built at increasing distances offshore. Moreover, it is not the case that conventional power plants are 100% reliable, and to deal with variations in the power available from conventional generators, and also with the variations in energy demand, the grid system is usually operated with some conventional plant on standby—it is kept running at low power as so called 'spinning reserve'. This can be used help compensate for any variations in supply from renewable sources, depending on the proportion of intermittent renewables involved.

In practice, so far, intermittency has not proved to be a problem in the UK. By 2002, the UK had around 3% of electricity coming from renewable sources, but the variations in supplies from the 500MW of wind farm capacity could not be detected by the grid controllers—the variations were lost in the normal variations in input power from conventional plants. The grid has to cope with potentially very large variations in input power, as, for example, if a conventional power plant trips out, suddenly removing up to 1GW of power input. The grid system also has to deal with occasional temporary failures of the cross channel link (1GW), as well as with major variations in demand patterns, driven by consumer behaviour, including their responses to TV schedules! So it is hardly surprising that the variations from renewables don't show up.[20]

However, when and if the total contribution from wind power to the national grid increases, the variable nature of the inputs to the grid may begin to have more impact. So far there do not seem to have been significant problems with the Danish system, which by 2002 was operating with an 18% contribution from wind. Indeed, at times, in some areas of the country, during periods of low demand (e.g. at night), most of the conventional plant was run down, leaving wind power as the main input. At other times, when there was more wind power than was needed, the excess energy has been used to provide heat to boost the accumulator heat store units associated with Combined Heat and Power plants, of which Denmark has many.[21]

Of course this does not mean that there could not be times when weather conditions reduce the net wind contribution considerably, and this will become more important as the percentage contribution from intermittent sources increases—above say 20%. There could then be a need for some extra standby plant, or for short term energy storage facilities, to maintain full grid security.

For the moment, there is no shortage of standby plant—most power systems operate with substantial 'plant margins' and with excess capacity over and above what would be needed to meet peak winter demand. As noted earlier, in 2002 the UK excess capacity was around 25%. But as more renewable energy capacity comes onto the grid in the decades ahead, at some point more standby capacity will have to be installed. Installing, maintaining and operating standby plant does cost money, although, at present, it is cheaper than providing energy storage. But, however it is achieved, dealing with intermittency need not be prohibitively expensive. In its UK Energy Review, published in 2002, the Cabinet Office's Performance and Innovation Unit concluded that the extra cost of dealing with inter-mittent sources like wind would be only 0.1p/kWh up to a 10% national contribution and 0.2p for 20%. Even with a 45% national contribution from intermittent sources, the extra cost would, they concluded, only be around 0.3p/kWh.[22]

In addition to the extra cost, the provision of back-up power would, of course, involve extra emissions, although some of the plants could use stored renewable fuel, like biomass. But, given that the periods for which they would be used would not be long, the use of fast-start-up gas turbines might be condoned, despite their carbon emissions. In some circumstances, energy storage can also be a viable option. Short term storage can be provided by a variety of small- to medium-scale mechanical and electric systems (e.g. as compressed air, via flywheels, in batteries) or by large-scale pumped storage schemes, although these techniques are all, to varying degrees, expensive. However, new electricity storage techniques are also emerging, such as the Regenesys system developed in the UK, which uses a REDOX chemical reaction. A 15MW prototype unit has been installed along side a gas-fired power station near Cambridge, designed as a grid back-up system, at a cost of around £20m, with subsequent versions expected to be cheaper.

System integration and the hydrogen option

For the immediate future, it seems that the existing power system can, in effect, act as back-up to compensate for intermittent inputs from wind, with the need for extra back up plant or energy storage being limited. In the medium term, given that the power system is also changing, there may be room for a significant increase in the contribution from intermittent renewables before we need to think about installing extra standby plants or energy storage systems. For example, the adoption of improved energy efficiency measures and demand management techniques can reduce peak loads, making it easier for the power system to cope with larger contributions from intermittent renewables. In the domestic sector, to deal with peak demand, interactive load management systems, such as 'smart plugs', can allow consumers to avoid periods of high power prices by switching off selected devices like freezers—which can happily run for a few hours without power.

In addition, there is increased reliance on power imports from neighbouring countries, which can help compensate for local intermittency in supply as well as local demand peaks. That seems to be one reason why the Danish system is operating successfully with 18% of wind on the grid—it can import balancing power from Scandinavia when and if needed e.g. from Norwegian hydro.

I have focussed so far just on wind. If other renewables, like wave power, are also feeding into the grid, the overall problem of intermittency may be lessened since the timing of the variations in power availability differ. For example, wave energy is, in effect, stored wind energy, gathered over a wide catchment area, and available long after the wind at any one point has dropped. In addition, energy from tidal flows, although cyclic, is unrelated to wind, and peak solar energy availability similarly depends on weather patterns not directly linking to wind patterns. With all these inputs feeding to the grid, the impact of variations in inputs from any one source could be less. One early study suggested that, assuming the provision of storage capacity, demand management and trading of power with neighbouring systems, an integrated power system using a range of intermittent renewables could be developed that was technically viable even if the contribution from variable renewables reached over 50% of the total energy used.[23]

Although there are still disagreements about the exact scale of the operational penalties and costs, it seems that intermittency is likely to be a technical integration problem, not a fundamental flaw, at least for moderate levels of variable inputs to a national grid system, from a range of intermittent renewables.

Fortunately, if we want to go beyond moderate levels of renewable contribution, there is a new, potentially cheap, longer term energy option opening up, which can help to balance the energy system—the use of hydrogen gas. Hydrogen can be generated, when power is available from intermittent renewables, by the electrolysis of water, and then stored for later use. Electrolysis provides a reasonably efficient way of using electricity to split water into hydrogen and oxygen: it has an energy conversion efficiency of up to 80%. The hydrogen gas could be stored or transmitted down conventional gas grids to where it was needed. In the interim, while natural gas is still available, it could be mixed with hydrogen. In fact, having some methane added in is actually important since with pure hydrogen there can be problems with embrittlement of gas pipes.[24]

In the longer term it could be that there could be a shift in emphasis from electricity transmission to hydrogen transmission. Gas distribution is much more efficient than electricity transmission, and gas can be stored whereas electricity cannot be easily stored. When burnt, hydrogen produces only water as a by-product. Although care has to be taken with this fuel, especially if stored as a liquid in cryogenic form, its use should not require much more in the way of safety provisions than we are used to with petrol, which after all is a very volatile fuel. In some ways hydrogen is safer than many conventional fuels. Being lighter than air, hydrogen gas rises, so if there are leaks then they vent out of houses, rather than collecting on the floor and building up until they reach a source of ignition, as can happen with conventional natural gas.

A switch over to what has been called 'the hydrogen economy' has been seen as a real possibility—as an alternative both to natural gas distribution and to electricity transmission. Ultimately, there could be giant hydrogen gas grids around the world, and hydrogen could also be tanked around, in cryogenic liquid form, in ships. Indeed this is already being done, with hydrogen from Canada's hydro plants being shipped to Sweden. In the longer term, countries in the Middle East might install large arrays of photovoltaic solar cells in desert areas so as to generate hydrogen to supply the less sunny areas of the world with a new clean fuel.

Hydrogen can be used as a clean fuel for vehicles or can be converted into electricity in a fuel cell. Fuel cell-powered vehicles are already emerging, and fuel cells are also increasingly being used for stationary power generation. Fuel cells can be highly efficient energy converters, with energy conversion efficiencies of up to 80%, especially if the heat produced is also used. It could be that in future there will be a global system based on renewable hydrogen, with electricity only being generated locally where and when needed. **See Box 5.**

Box 5: Iceland's Hydrogen Economy

Iceland is currently the largest user of renewables, with around 99% of electricity coming via geothermal geysers and hydroelectric dams, but it has only explored about 14% of this potential and it is committed to use its renewable resources to become the world's first hydrogen economy, hoping to cut greenhouse emissions to zero within 30 years. At present, it imports oil to meet 35% of its energy needs (including transport), and this makes the country one of the world's higher per capita carbon emitters, but, on the basis of a programme being backed by DaimlerChrysler, Shell Hydrogen and Norsk Hydro, in a joint Icelandic New Energy venture, the aim is to lead the conversion into a hydrogen economy by following a six-step plan which could take 30 years to complete. The first phase, now running, is a £4m programme, subsidised by the EU, to run a trial on three hydrogen buses and to add a hydrogen station in a conventional petrol station. The second phase will convert the island's buses to hydrogen, followed by all cars. The fourth and fifth phases should convert the fishing fleet. The sixth and most adventurous phase is to export hydrogen to Europe. Source: *The Guardian*, 18 July 2001.

Chapter 2

Renewable energy: its limits and opportunities

The problems ahead

So, having set the scene with an outline of how things might be, let's recap. It does look as if there is a technically viable set of possibilities available based on the use of renewable energy—although of course there is a long way to go before we could get to the sort of future described above. But it does seem that, at long last, renewable energy is about to lift off in the UK. The government's White Paper on Energy published in February 2003 talked of an aspiration to obtain 20% of UK electricity from renewables by 2020.[25]

It is common, when analysing why it has taken so long to get stared with renewables, to point to the various obstacles thrown up by those with vested interests in the current energy system—for example the economically and politically dominant fossil fuel and nuclear interests. I could fill this whole Briefing with examples. The nuclear industry could maybe have a separate edition of its own—it has cornered the lion's share of any funding that was going for new energy technologies for decades. But instead of looking backwards, I want to assume that these various institutional obstacles can and will be overcome, as a new green energy paradigm becomes established. That's a big assumption of course, but there does seem to be growing support for change, even in the most unlikely places, as witness Shell and BP's partial conversion to sustainable energy.

Inevitably, there will still be problems ahead dealing with those who want to look backward rather than forward, but I want to deal with some of the practical and strategic problems that will face us as we try to move to a sustainable energy future. For as I hope to show, the way ahead is not as simple or trouble-free as we might hope—and we need to recognise the problems in order to deal with them.

Otherwise, those who still oppose the switch over will use them to oppose change.

I am going to look at three areas, concerning respectively short, medium and long term problems. The short-term example will concern the fact that no technology is entirely benign. Renewable energy technologies like wind farms may have far less impact than the global impact of fossil-fuelled plants, but they do have some local impacts, in terms, for example, of visual intrusion. Some people feel we must protect treasured views or unspoilt areas at all cost—but what if the cost is the collapse of the wider ecosystem? How do we trade off local and global impacts? To try to answer that I will be looking at the UK wind farm issue and at some other examples—including the problems facing hydro power and waste combustion.

My medium term example concerns transport. It is fairly easy to outline a sustainable energy path for providing electricity and heat, but what about transport fuels? Can we develop sufficient green energy systems to meet that demand as well—especially given that it is increasing as more and more people want mobility. Here we come up against the limits of technology as a way of resolving environmental problems. Or to put it another way, we move on to having to look at changes in the way we live and behave.

That links in to my final example, concerning the longer term limits of the sustainable energy approach. Is it possible to use renewable sources to sustain economic growth indefinitely? What are the limits?

Renewable energy—the options

Renewable energy sources could meet a substantial part of our energy requirements. A scenario produced by Shell in 1995 suggested that by 2060 renewables could supply about half the world's total energy needs.[26] By 2100 we might even be able to obtain more or less all of our energy from renewables, should we wish. So far only relatively slow progress has been made in this direction. The European Union's target is to obtain 12.5 % of its electricity from the so-called 'new' renewables, i.e. leaving aside the well-established hydro plants, by the year 2010. The UK's target is 10% of electricity by 2010, and 20% by 2020.

The economics of renewables are improving rapidly. Wind power costs have fallen dramatically as new technology has emerged and as operating experience has been gained. That has built investor confidence and allowed for more favourable financial arrangements. The first projects in the UK had to be given 11p/kWh, under the Non-Fossil Fuel subsidy system, whereas by 2000 projects were going ahead, on some good sites, at around 2p/kWh, comparable with gas-fired plants. Most of the other renewables look likely to follow similar patterns of price reduction. A review carried out by the UK Cabinet Office Performance and Innovation Unit produced the figures in Table 1 for prices by 2020, illustrating that most renewables could be competitive with most conventional energy options.

As can be seen, on-land wind was assessed as being likely to be the cheapest option—even cheaper than combined cycle gas turbines. That is hardly surprising since it is already an established technology. In principle on-land wind might supply up to 10% of UK electricity. Offshore wind is, as yet, less developed, but it too looks very attractive economically and the energy potential is very large—at least 20% of UK electricity and probably much more, depending how far out to sea we can go.

Table 1: **Cost of Electricity in the UK in 2020**

	pence/kWh
On Land wind	1.5–2.5
Offshore wind	2–3
Energy crops	2.5–4
Wave and tidal power	3–6
PV Solar	10–16
Gas CCGT	2–2.3
Large CHP/cogeneration	under 2p
Micro CHP	2.3–3.5
Coal (IGCC)	3–3.5
Nuclear	3–4

Source: Performance and Innovation Unit, 'The Energy Review', UK Cabinet Office, 2002 http://www.piu.gov.uk/2002/energy/report/index.htm

The newer renewables—energy crops, offshore wave and tidal current power—also have large energy potentials but are seen, at least by the PIU, as likely to remain relatively expensive, while the PIU sees solar PV as likely to remain very expensive even by 2020. However, not everyone would agree with these high estimates. Better progress may be made as the technology develops. For example, there are several tidal current projects under test which are claimed to be likely to lead to full scale generators able to supply electricity at around 4p/kWh within a few years. One device is the Seaflow marine turbine, a 300kW prototype of which is being tested off the coast of Devon near Lynmouth.

Meanwhile in the UK's Severn estuary, Tidal Hydraulic Generator Ltd are about to test five of their underwater turbines, which are to be placed on the estuary floor between the two bridges. Going north, a novel hydroplane device, known as the Stingray, has been tested off the Shetlands, and in Norway, ABB and Statoil are testing their own tidal current system. And over in San Francisco, a novel UK-designed tidal current system is to be tested on the seabed near the Golden Gate bridge.

Similar developments are happening in the wave energy field— in the UK, Denmark, the Netherlands, Sweden, Japan and elsewhere. It's an exciting time to be involved with these new marine renewable technologies—after so many years of being starved of funds, at last some money is beginning to flow. And as it does, the technologies are likely to develop very rapidly, so that prices fall.

The same is true for PV solar. For example, an update report was produced in November 2002 by Imperial College's Centre for Energy Policy and Technology, at the request of the Prime Minister's Strategy Unit (which is what the PIU is now called). It quoted likely grid-linked solar PV generation costs 'beyond 2020' as being around 8 US cents/kWh in the UK, 5 cents in southern Europe and 4 cents in most developing countries.

Certainly, as noted earlier, continued technological breakthroughs are likely to lead to cheaper and more efficient cell materials. In parallel there are novel applications which can compensate for the relatively high cost of existing cells. For example, if we use PV materials, like the new 'solar tiles', which look just like ordinary roof slates, instead of conventional roofing materials, then the overall cost is reduced, since you don't have to buy roofing materials.

Building-Integrated Solar roofing and façades are becoming common for corporate headquarters—they are cheaper than some of the prestige cladding materials, like the marble façades often used. And they also generate power! Niche markets like this can help build demand for PV which, in turn, helps reduce the price.

Of course, simple economic analysis may disguise other benefits. For example, energy crop-based systems may still be relatively expensive, but it must surely make more sense to help farmers to develop energy crop plantations if the alternative is just to pay them keep their land unused as 'set aside'. Given time, it looks as if energy crops could become a major source of renewable energy in the UK, as well as elsewhere.

For the moment though, on-land wind power is clearly the most obviously viable renewable energy option. Around the world, as already noted, about 32,000 megawatts of wind generating plant has been installed so far, 12,000MW of it in Germany. However, the UK is trailing far behind with just over 500MW. That has involved finding sites for 1000 turbines and, in some areas it has become increasingly hard to find more sites. Around 70% of applications are turned down due to local environmental objections. This first case study looks at the reasons why, and at what might be done about it.

Wind power—no thanks

The UK wind farm programme, which got underway from 1990 onwards, owes its existence primarily to the 'non-fossil fuel' cross-subsidy scheme introduced following the privatisation of the electricity supply industry in that year.

The 'non-fossil fuel obligation' (NFFO) was imposed on the newly privatised regional electricity supply companies, requiring them to buy in set amounts of electricity from nuclear and, to a much lesser extent, renewable suppliers. In addition, a surcharge was imposed on fossil fuel electricity generation in order to meet the extra cost of using non-fossil sources, this cost, ultimately, being passed on by the energy supply companies to their customers.

Between 1990–1998, five NFFO orders were set for renewables, offering selected projects a 'premium' price, over and above the

usual 'pool' price, for their electricity. Most of the initial series of wind farm projects supported under the first NFFO got through, and, although objections began to emerge subsequently, 54 MW of the 84 MW of wind capacity contracted for under the second NFFO has been successfully commissioned and 11 wind farm projects supported under the third round of the NFFO obtained planning permission, while only 7 were turned down.

However since then the rate of refusals has increased, so that by the late 1990s nearly 70% of the projects failed to get planning permission. Most of the objections have concerned visual intrusion. Between September 1991 and December 1993, 12 wind farm proposals went before planning inquiries and 9 of these were approved. But between January 1994 and January 1999, only 2 of the 18 proposals called in for decision by the minister, won approval following a planning inquiry.

Although only a proportion of wind farm proposals are called in for planning inquiries, even so, the trend seems clear. Dr Peter Musgrove, from National Wind Power, told the *Times* on 9 January 1999: 'Since 1994, planners and inquiry inspectors have been giving progressively less weight to the clean energy benefits of wind farms and progressively more to their negative and subjective assessment of visual impact'.

As the *Times* put it, 'Unless urgent action is taken many firms will leave Britain for windpower opportunities overseas', and the

UK would find it hard to meet the target of obtaining 10% of its electricity from renewabies by 2010.

Standing back, there is something of a contradiction in the situation, in that there is broad support for wind farms in the UK, and local support actually seems to grow once wind farms have been established. For example, a study on people living near Scotland's first four wind farms indicated that 40% had felt there would be a problem but only 9% reported any. Overall, two-thirds of those asked found 'something they liked' about wind farms. Moreover, a poll carried out by the British Market Research Bureau for the Royal Society for the Protection of Birds in 2001, found that only 3% of those asked objected to wind farms. However, there is also some strong local opposition, and, even if that is in a minority, it seems to be slowing the rate of deployment dramatically.

Solutions

What might be done to resolve this problem? Firstly we must recognise that not all wind projects are necessarily well thought out. Certainly there are a variety of practical steps that could be taken to improve the situation, such as better location of wind farms, and the use of waste land or old industrial sites. Technological developments may change the situation. The introduction of more efficient variable speed wind turbines, which can operate effectively with lower wind speeds, could reduce pressure on upland sites. As we have seen, costs have also dropped dramatically since the early days, to one third of the cost of wind projects established in 1990. That makes it more economically viable to choose less windy sites. However, to ensure that these technical and economic improvements translate into environmental/locational improvements will probably require something more coherent than the UK's current rather ad hoc approach to planning in relation to wind projects, which operates on a site-by-site basis.

Obviously it is sensible to assess each project on its merits, with local issues and concerns being taken into account. This can lead to expensive planning delays and one solution would be to develop some form of zoning, with areas suitable for development identified in advance, as in Denmark. Something like this is in hand in England with the development of a regionally-based planning framework and local councils setting targets for the amount of renewable energy to be obtained in their regions.

There is also an emphasis in the UK towards streamlining the planning system to avoid long delays and provision for Parliament rather than local inquiries to determine some major infrastructure projects. However, if that led to any weakening of planning control it would be likely to be seen as reprehensible by most environmental groups, even by those in favour of wind, since it might open up the way for less desirable projects.

What about new patterns of ownership? Most of the wind projects are owned and controlled, one way or another, by a few large companies, and this has sometimes been a source of resentment. For example, one of the issues raised during the local campaign against the 39 turbine wind farm given planning permission for a site at Cefn Croes in mid-Wales was that it would will not benefit local people and was being 'imposed on the community by Enron Wind, a subsidiary of the shamed US multinational' (*The Guardian*, 20 February 2002).

By contrast, opposition to wind farms has been far less apparent in Denmark and Germany, where the majority of wind projects are owned by local people, who share in the profits. The growth in privately owned as opposed to utility owned projects in Denmark has been quite striking. Around 80% of 300MW of new wind capacity installed in 1997 involved projects owned by individuals or local wind co-ops, called Wind Guilds. There were changes in the subsidy system in the mid 1990s which slowed the rate of growth of locally owned projects, but even so, by 2000, 75% of the 900MW then installed was owned by local co-ops or individuals, in roughly equal numbers, and half of the large 40MW offshore wind farm, installed in 2001 about 2km off the coast from Copenhagen, is owned by 9,000 local co-op members. Overall, more than 100,000 families own shares in local wind projects in Denmark, out of a population of 5 million.

Local ownership via community-based co-ops (known as Windmill Associations) is also popular in the Netherlands. Similarly, in Germany, local interests dominate ownership, with around two-thirds of the projects being owned/operated by local farmers, homeowners and small businesses—and, as we have seen, by 2002,

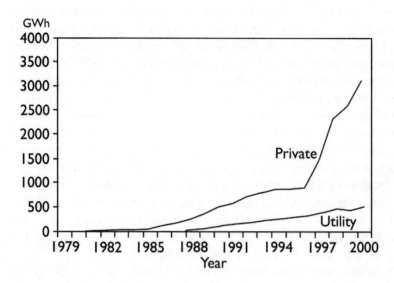

Figure 1: **Growth of Local ownership of wind projects in Denmark**
Source: Data from Paul Gipe

Germany had installed around 12,000 MW of wind capacity, compared with the UK's 500 MW. While in the UK some people have been clamouring for wind projects to be halted, in much of the rest of Europe many people seem to be clamouring to be able to benefit economically.

If wind projects involve some form of local control and local economic benefits, the local impacts may, it seems, be judged more favourably. Danish enthusiasts for local ownership of wind projects often repeat the old Danish proverb: 'Your own pigs don't smell.'

Put more positively, to the extent that the problem is local residents' sense of resentment in the face of what they perceive to be projects imposed on them by outsiders, who profit from them, then the solution is for members of the local community to get directly involved, via ownership. That can lead to direct economic benefits as well as to more indirect gains, such as local economic regeneration. For example, farmers, who are finding it hard to survive economically by producing food, might diversify in to wind power and other renewable projects, thus strengthening the local economy.

So far, there are only two Wind Co-ops in the UK—Baywind in Cumbria and Cwmni Gwynt Teg in Wales—although some other initiatives are underway. One Welsh group, Awel Aman Tawe, has taken pains to consult widely with local people, and in 2001 it organised a public referendum in the area to decide whether a wind co-op project should go ahead (the result was 57.5% for, 42.5% against).

Of course, even with conventional commercial projects, there are some local benefits, in terms of employment during construction and tourism subsequently: some sites have set up visitor centres which have proved popular tourist attractions, injecting cash into the local service economy. Local farmers can also benefit, by renting out fields for use by wind farm developers (annual rentals of £1,000-£2,000 per acre or per turbine are typical), although this can sometimes lead to local conflicts and resentment if residents feel they are adversely effected by the wind farm.

In some cases wind farm developers have offered packages of 'community support measures'. For example, National Wind Power provided £100,000 for a Charitable Trust to be used to support local schools, colleges, students, apprentices and training schemes in the area around its wind farm at Bryntitli in mid-Wales. It has also provided a £5,000 p.a. fund to support local environmental improvements in the area of the wind farm, and has set up a Community Fund which will receive £5,000 per annum for the benefit of local inhabitants. Schemes like this might give local communities a vested interest in the project, or failing that, might be seen as compensation for any disbenefits, although, equally, they might be seen by opponents of the project as a form of bribery.

Another form of involvement, albeit rather diffuse, can occur if consumers contract with one of the dozen or more green power retail schemes that are on offer. In these schemes, electricity suppliers promise to match the power consumers use with power bought in from renewable sources. In some schemes a small surcharge is made. By 2002 around 60,000 people had signed up in the UK.[27]

Some of the schemes use wind power, and participating consumers are able to identify the sites used. This may not be ownership, and it may not be a local plant, but it does make some sort of link between what you get out of the socket in the wall, and the source of the power. Moreover, in some schemes, consumers actually donate a surcharge direct to a fund or trust which invests in specific renewable energy projects, usually locally. So in that case there is an even more direct sense of local involvement.

There are also other changes that are occurring in the electricity market, which might provide an opportunity to help improve the situation. Before liberalisation of the electricity market, everyone in the UK paid the same for electricity regardless of where they lived. Now, however, consumers may have to pay different amounts depending where they live and who they buy their power from. Before liberalisation, people in remote areas were, in effect, subsidised by those who lived nearer to power stations—since there are power losses associated with long distance transmission. But, if power is produced locally, it is clearly valuable to use it to meet local energy needs. In which case, should not people who live near wind farms be charged less than those at a distance? That would compensate directly for any loss of local amenity. However, this idea opens up the possibility that other groups in society might seek to claim compensation—for living next to a nuclear plant for example. But it would establish some sort of link between costs and benefits.

Social control of technology

'I suspect that future generations will look back at controversy over wind farm sites and wonder what all the fuss was about. They will be part of our landscape and seascape and we will be pleased to have them. Although it is important to take local wishes into account, we should not allow a NIMBYist stance to impose a veto on onshore wind farm developments.' So said Peter Hain, then Energy Minister, during a House of Commons debate on Renewable Energy in 2001.

It is certainly interesting that there were tens of thousands of traditional windmills dotted around the UK until the eighteenth century and the survivors are now cherished as beautiful. However it is perhaps too easy to depict objectors to modern wind turbines as simply responding to NIMBY sentiments. The debate can also be seen as reflecting an important concern for environmental protection. As an opponent of a Scottish wind farm put it, 'We are the custodians of this landscape for future generations and as such it is our duty to fight this alien intrusion.'

One of the fundamental issues that emerges from the wind case study relates to people's sense of lack of power to protect their interests, which may include altruistic environmental concerns. However, in some cases, the development of environmental sensitivity can lead to some contradictions when it comes to choosing the focus for specific campaigns. For example, if you are concerned to protect your local environment, you can seemingly have far more effect by opposing a local wind farm or some other project in their area, than by campaigning to halt global climate change. The former can be seen as an immediate and visible threat, the latter is remote, longer term and, in any case, seemingly beyond your influence. The solution seems to be to increase the opportunity for involvement with and benefit from such projects.

That does not mean we have to accept all the projects that are proposed uncritically. For example, waste-into-energy plants involving power production from the combustion of domestic and industrial waste seem to have serious potential problems. Certainly some of these projects have met with strong local opposition from residents concerned about toxic emissions and unconvinced by assurances that these would be kept within controlled limits. National level environmental groups also opposed waste combustion. Greenpeace mounted a very successful 'Ban the Burn' campaign, labelling waste combustion plants as 'cancer factories'. In addition to the issue of emissions, groups like Friends of the Earth felt that energy from waste combustion, as a profitable waste disposal option, would undermine efforts to increase the scale of waste recycling and the development of a more sustainable approach to waste management. For example, given that there are economies of scale with waste combustion plants, large plants are favoured, and there have evidently been occasions when waste has had to be transported long distances, even from other cities, in order to keep the plants fed with fuel.

Given local opposition, many of the waste combustion projects offered support under the Non-Fossil Fuel Obligation (NFFO) could not be completed: only around 72 MW of the 311 MW of municipal and industrial waste combustion capacity supported under the first two NFFO rounds was successfully commissioned. Subsequently, conventional 'mass burn' combustion projects were excluded from the new support system that has replaced the NFFO, the Renewables Obligation, ostensibly since they were deemed to be commercially viable on an independent basis so that they no longer needed extra support. Instead the UK government called for a focus on what they saw as potentially cleaner waste processing technologies—pyrolysis and gasification—which were included as eligible for support under the Renewables Obligation.

It is not yet clear whether these new energy recovery technologies will win favour with environmentalists—the technologies are, as yet, relatively underdeveloped. But it is clear that most environmentalists feel that the case against conventional 'mass burn' waste combustion is very strong—it is not seen as a genuine or safe renewable energy source.

What we are seeing here is an ongoing process of technology assessment—a process being carried out as new technologies emerge, with the involvement of local people. Sometimes the results may initially be contradictory, as with wind power, where a minority are currently opposed. But rather than seeing that as a problem, we should perhaps welcome it as opening up a debate about which sort of technologies we should focus on.

Some wind farm objectors have tried to adopt a more positive approach by pointing to alternative possibilities—for example offshore wind farms. Certainly, the UK's offshore wind resource is very large, and the environmental impacts associated with tapping them are generally low. But, given the UK's 10% by 2010 renewable energy target, offshore wind is likely to be in additional to, rather than instead of, on-land siting. Moreover, offshore siting is more expensive than on-shore wind, primarily due to the need to transmit power by undersea cable to the shore, and, unless offshore wind projects are sited well out to sea (thus increasing cost further), there could still be objections about visual intrusion. Indeed in 2002, there were strong local objections to the offshore wind farm proposed for a sandbank 5 miles out from the Scottish side of Solway Firth, and a local campaign against the project proposed off the coast of south Wales near Porthcawl.

Energy conservation is obviously another very important alternative, but, as we will be discussing later, there are limits to what it can achieve. For example, once the opportunities for cheap and easy savings have been exhausted, the cost effectiveness of energy saving measures is not much different, in terms of carbon emissions avoided per pound spent, than that obtained from wind power projects.

Moreover, if the climate change problem is as serious as many think, and we wish to avoid the nuclear option, then we will need all the energy conservation and all the renewable sources we can reasonably muster—it is not a matter of either/or. This conclusion becomes even more stark if we assume, as seems likely, that, even given a major commitment to energy conservation, overall demand for power will continue to rise, given the ever growing lifestyle expectations of our consumerist society.

Of course this situation is not unchangeable. As I will be discussing later, we may be able to move to more sustainable forms of consumption. However that could require major social changes, and, if it was to be in any way equitable, there would be a need for a major exercise in negotiation amongst the various competing interest groups in society. Be that as it may, for the moment it seems that most people cannot accept life without more and more energy-using devices, from dishwashers to tumble dryers. In which case the point has to be made that the energy will have to come from somewhere. If we want to avoid the problems of climate change and other environmental problems associated with the conventional energy technologies, including nuclear power, then we will have to develop as many of the renewable alternatives as we can.

This obviously has implications well beyond the UK and indeed well beyond renewable energy. I will be looking at some of the wider issues concerning the limits to material growth and rising economic expectations later. But for the moment, let's just limit ourselves to some of the technological choices facing the developing world, and in particular the problems raised by one of the renewable energy options: large hydro.

Energy and development

Many developing countries have traditionally seen the installation of large hydropower projects as part of the process of modernisation, and companies from the industrialised countries have been more than happy to oblige. Large hydro is a major energy supplier—there is around 650 GW of installed capacity in place, mostly in 300 large projects.

However, in recent years, there have been social and environmental concerns about large hydro, and some new projects have met with opposition. One key issue has been the problem of social dislocation caused by having to move people from the areas being inundated to create reservoirs. In addition, it seems that in some locations (e.g. in warm climates), the anaerobic digestion of biomass brought continually downstream and tapped by the dam can create methane gas to such an extent that a coal-fired plant of the same capacity would produce less net greenhouse gas impact.[28]

What we are seeing here is the result of the disruption of a natural energy flow, which previously ensured the continual agitation of the water, so that anaerobic processes were minimized. There may be solutions, including the redesign of dams so that some water flows are redirected to continually stir up the 'sump' in front of the dam, and the scale of this problem will in any case depend on the local topography and ecology.

While some large hydro may have problems, smaller-scale hydro—as in so called 'mini' or 'micro hydro' projects in rivers and streams—continue to be developed successfully around the world. There are 50,000 small hydropower stations in China alone. The individual generation capacities involved are small, but taken together they represent a significant amount of power, and this option is particularly relevant in developing countries, providing cheap local power inputs. Micro hydro also avoids most of the environmental problems associated with massive hydro schemes. Certainly small projects are less locally invasive.

There are many other small- and medium-scale energy options available for developing countries, most obviously solar energy. Indeed, solar photovoltaic cells may be the only hope that many of the 2 billion or so people in the world currently without electricity have to get access to it. Biomass and biogas production are also obviously very well suited to many areas in the developing world, and this can be done on a small local scale—but using the latest technology. For example, one Indian project involves the production of hydrogen gas from biomass using bacteria. The Chettiar Research Centre has explored hydrogen production from whey, sago, sugar effluent and distillery waste, and has been testing the use of biologically produced hydrogen in a 250W fuel cell for electric power generation.

Moving upscale there is wind power. India is one of the front runners in the wind power race, with over 1.5 GW installed, while China is expected to have around 3GW of wind capacity installed by 2005 and perhaps 5GW by 2010. And if we really must have large systems then one option is tidal current power—taking energy from tidal flows. The Canadian company Blue Energy is developing an ambitious 'tidal fence', with an array of vertical axis water turbines mounted in a causeway between islands in the Philippines, which could ultimately be expanded to deliver 2,200 MW at full power. That would have a much smaller environmental impact than an equivalent capacity hydro project—there would be no need to dam rivers and flood land.

Clearly, in each region of the world different patterns of renewable energy use will emerge, and there will have to be a process of negotiation and debate about which technologies are most appropriate, in terms of scale, impact and location. Smaller-scale projects have the attraction that they can be more easily financed and, hopefully, will relate more easily to the local social context, in some cases opening up the opportunity for locally owned and managed community projects. By contrast, large projects will have to involve national level management and finance, and often the involvement international companies and financial backers.

It is clear that if renewable energy systems are to diffuse across the world, many of the less developed countries will need technical and financial support. The Kyoto Climate Change Accord includes a commitment to a Clean Development Mechanism designed to stimulate 'technology transfer' of low carbon energy systems from industrialised countries, i.e. projects installed by companies from the developed world in developing countries. In return the developers will receive 'carbon credits' for the emissions avoided, which can be used to set aside emissions back home, or for trading in what seems likely to be an expanding emission credit market. Nuclear power has, so far, been excluded from this arrangement. But interestingly, there has also been strong opposition to the idea of including large hydro projects in this scheme, in the belief not only that it was an inappropriate technology, but also since large projects like this might 'drown out' the smaller, more locally appropriate schemes.

Living with renewables

As we have seen, not all renewables are equally well received or equally acceptable, depending on the context. However, it is early days yet for this new area of technology, and views as to which technologies should be developed, and in what ways, are still being formed. That is not surprising. Renewable energy effectively represents a new form of energy technology, involving new patterns of energy production and use, with new types of social and environmental impact. Whereas previously the emphasis has been on concentrated energy sources and centralised power plants, the trend seems to be towards the more decentralised use of diffuse natural energy flows and sources.

The emphasis would thus be on natural processes occurring in 'real time', not with inherited wealth from stored fossil or fissile energy. That opens up a whole new range of planning and land use issues, which have only just begun to be discussed.

While some renewable energy enthusiasts have expressed concern at the tone of some of the current debate in the UK over the merits of wind farms, and some of the battles over large-scale hydro in the developing world have become very polarised, in general, public debate should be welcomed, as long as it is as well informed as possible. Renewable energy technology is meant to be both socially and environmentally acceptable. Certainly in principle, the development of renewable energy should be more amenable to local social control, since many of the technologies are relatively small scale, and their nature, function and likely impacts are relatively easy to understand.

However, the choices are not always simple. For example, in some cases large-scale land-using options like large tidal barrages may be seen as undesirable, compared say to the much less environmentally invasive option of exploiting tidal currents using free-standing turbines on estuary beds. The same may be true of large hydro projects. However, there can also be aspects of large-scale tidal or hydro projects which could in fact prove socially and environmentally attractive, quite apart from non-fossil energy production—for example the provision of new habitats for new species to compensate for those lost by inundation, the creation of new water-based leisure facilities for water sports and recreation, and the increased availability of drinking water from the reservoirs that are produced.

Although in the case of large hydro and tidal barrages the negative impacts may not always fully compensate for secondary benefits like this, there are other examples where the secondary benefits could in fact become a dominant concern. For example, the UK government recently identified three specific areas off the coast of the UK as being suitable for wind farm development and has earmarked these zones for strategic assessment. As it happens, these overlap quite well with cod spawning and nursery areas. It could be that arrays of sea-bed-mounted wind turbines out to sea could provide new habitats for marine life—it is well documented that crustaceans and cod find wrecks and other man-made intrusions attractive, as long as they do not introduce toxic materials into the environment. In addition, a network of wind turbines out to sea could provide a base for fishery protection activities, with electronic devices being located on the turbines to monitor the access of trawlers. The same could be true for offshore wave energy farms and tidal current farms. In addition, wave and tidal energy extraction systems could help reduce coastal erosion in some areas. Looking to the future, it might even be that some parts of the North Sea near to the coasts could have an interlocking network of man-made causeways providing a base for wind, wave and tidal devices, as well as sea defences and fishery protection services, and also, possibly, new opportunities for leisure pursuits.[29]

As these ideas suggest, at the very least the adoption of renewables may open up not only new opportunities, but also a wide range of planning and environmental issues. They may also open up a range of sometimes quite difficult strategic issues, not least in terms of which renewables should be adopted in developing countries. The debate over large hydro has already been mentioned—it is usually argued that smaller-scale renewables would be much more appropriate both socially and environmentally. However, not everyone agrees with the idea, promoted by some environmental groups and aid agencies, of pushing ahead with the early, widespread and rapid adoption of solar PV as an aid to development. Some critics see PV as too expensive at present, and argue that only when costs are reduced by development and deployment in the industrial countries, can PV be usefully deployed on a wide scale elsewhere. For the moment, it is sometimes argued, in much of the developing world, the use of biomass, small-scale hydro and possibly small-scale wind makes more sense.[30]

Clearly, as well as there being some interesting opportunities, there are some difficult choices ahead, and consequently a need for public debate over specific projects and over longer term development patterns locally, nationally and internationally. However, debates (and potentially disagreements) should be seen not as a problem, but as an opportunity to increase public participation in shaping the future, and as part of the process of bringing technology under more direct social control. There is no one road ahead—we need to negotiate the best path to the future.

Chapter 3
Transport: the hard road ahead

The limits to mobility

Even given the constraints and problems we have discussed, the way ahead for renewables does not seem too difficult—as long as the conflicts of interests can be properly negotiated. Certainly it is relatively easy, in technological terms, to devise energy supply strategies to deal with electricity supply and also heat supply. However, when it comes to transport, the problems of matching supply to demand get harder. Demand for mobility is increasing worldwide as affluence spreads, with private cars being the vehicle of choice. Worse still, the rate of this increase in demand is significantly higher than for energy for other purposes. **See Fig. 2** for the UK situation. As can be seen, while industrial energy use has been cut successfully, demand in the domestic and service sectors is still growing, and in the transport sector it has been roaring upward. The recent economic slowdown may have led to a fall in transport

Final energy consumption, 1970 to 2001

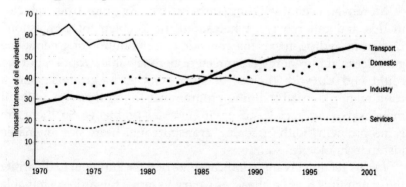

Figure 2: **Energy Consumption by sector in the UK**
Source: data from 'UK Energy in Brief', Department of Trade and Industry, 2002.

energy use in the UK, as the graph shows post-2000, but this is likely to be temporary. As the global economy expands, so too will transport demand.

Can the situation be rescued by switching over to non-fossil fuels? In energy terms, ultimately we might in theory be able to meet demand for personal mobility from renewable sources—gaseous and liquid fuels produced from biomass, hydrogen generated by the electrolysis of water using power from wind farms, wave energy or tidal projects. But that would be at the expense of other uses of these green sources, and, in the case of biomass, other uses for the land. At present, in the EU, large areas have been 'set aside' from food production to protect the profits of the farming industry, but even if this cannot be changed, growing fuel for cars may not be the best use for land.

One option sometimes put forward is growing trees to absorb the carbon dioxide emitted from cars and from power stations. Obviously reafforestation has its attractions as a carbon store, since it is relatively cheap, assuming land is available, and if it is done right it can offer other benefits, such as enhanced biodiversity. However, the problem is that the areas needed would be vast to make any impact on carbon dioxide emissions. In addition, the storage is not permanent. Forests can burn and trees eventually decay, releasing the stored carbon dioxide, so you would have to replant regularly. It has been estimated that to store all the UK's continuing carbon emissions in trees would require new forests to be planted over an area the size of Devon and Cornwall every year.[31]

Moreover, even if we can reduce carbon emissions from vehicles in this and other ways, what about the problem of congestion? Electric or hydrogen cars may not add to air pollution or greenhouse gas release, if the energy is sourced from renewable supplies, but we could end up with queues of such vehicles in gridlock, and a continuation of road building programmes to put them on, with all their associated social and environmental problems. So there are some problems with 'greening' transport which are not amenable just to technological solutions.

In this section I want to look at some of these problem to illustrate the constraints that lie ahead as we try to move towards sustainable approaches to energy use.

Hydrogen drivers

The first problem identified above concerned competing uses for renewables. To put it starkly, if we want to try to replace coal burning and nuclear power with renewables, and also begin to replace gas burning for electricity and for heating with renewables, then there is not going to be enough renewable energy to also take on transport— at least, not for some while. All may not be lost, however. There is another way to reduce emissions from transport, at least for the medium term, even if we may not like some aspects of it.

BP, Shell and the like would like to see cars running on hydrogen produced from fossil fuels. The conversion to hydrogen would be carried out in large plants and the carbon dioxide gas produced by the conversion process would be collected and then sequestered (i.e. stored) in old, empty, oil and gas wells. With this option in mind, the Shell energy scenarios produced in 2001 revised their earlier view, which, as I noted in Chapter Two, suggested that the 'new' renewables (i.e. excluding hydro and traditional biomass) could contribute up to 50% of world energy by 2060. In their new scenarios, what they seem much more interested in is devising a strategy for continued expansion of the use of key fossil fuels, with some hydrocarbons being used to produce hydrogen gas, and with carbon dioxide sequestration being used to avoid the climate change impact. On this basis, in Shell's 2001 'Spirit of the Coming Age' scenario, new renewables only need to expand to make a 22% primary energy contribution by 2050.[32]

The idea seems to be catching on. In February 2003 no less than US President Bush lent his support to a $1.2billion programme to push the hydrogen option on—with coal being one of his favoured sources. Must we accept the continued use of fossil fuels in this way since we can't hope to produce sufficient renewable energy, at least for a while?

Certainly it seems that way. Renewables look like they will mainly try to compete in the electricity market. For example, they seem likely, given support, to be able to expand up to 20% or even 30% of electricity supply in the UK by 2020, and possibly more if we really tried hard. So that replaces nuclear and most of the remaining coal, much of which is imported. Solar and biomass heating may offset some gas used for heating, as may the use of heat pumps and micro Combined Heat and Power units (these latter two use gas more efficiently). Some biomass and wastes may be used to produce transport fuels—liquids like green diesel or gases like methane and possibly hydrogen. But given the land-use limitations, the contribution from biofuels to the transport sector cannot be very large in crowded countries like the UK—just for niche markets such as inner city hydrogen taxis, rural agricultural vehicles, fleet vehicles, and some buses.

Given that vehicle fuels command high prices, there will be a temptation to use any biomass available to feed this market. This could be sensible, for example for urban wastes. Indeed some urban wastes can be readily used for this purpose—used restaurant and chip shop cooking fat for example. It might also make sense to use some agricultural wastes similarly, although most of them would probably be better used for on-farm heating (e.g. straw burners) and biogas production—or for combustion in electricity generating plants, as is now being done with pig slurry and chicken litter. Certainly in terms of specially grown biofuels, the land use constraints would suggest that it would be better to use scarce land more efficiently to produce solid energy crops for electricity production—the energy output to input ratios are much higher for wood chips from short-rotation willow coppice than for liquid rape methyl ester from oilseed rape.

The situation in some other parts of the world may of course be different. For example, as we saw in Box 5, Iceland has enough renewable sources to begin to take on transport now. On a larger scale, in some countries where land availability is not a major constraint, it may be reasonable to use it for growing transport fuels, although even so there is still the risk that, for some developing countries, biofuels for vehicles would become a new 'cash crop' for export and, if carried out too extensively under the impact of global market demand, that could undermine biodiversity.

Leaving that issue aside for the immediate future, it seems that within the industrial world there may not be much of a showing from renewables in the transport sector until maybe 2020. That will leave the field open to hydrogen from fossil fuels, assuming that the transport sector does try to mend its ways and move away from the use of petrol. A switch over to fossil-derived hydrogen may not be too bad, assuming it involves sequestration of the emissions, since it will open up the hydrogen economy—with demand from the transport sector being a strong driver pushing hydrogen forward, ready for renewable sources of hydrogen to take over when they become more available.

A switch to hydrogen from fossil fuels as an interim measure also ties in reasonably well with how much space there is for storing carbon dioxide gas in empty oil and gas wells. For example it has been estimated that oil fields, globally, could take from 40 to 100 Gtonnes of carbon (Gtc), while gas fields might take 90-400 gtc. The ranges indicate just how uncertain we are about exactly what might be available. For comparison the global CO_2 emission from fossil fuel combustion was 6 Gtc in 1990, so we are only talking about a storage capacity of at best 80 years and maybe much less.[33]

Using old gas and oil wells seems reasonably safe—they held gas and/or oil for millions of years, so replacing that with carbon dioxide, even under pressure, should not be too risky in term of leakages. The safety and security problems associated with using the much large volumes that are in theory available in aquifers are much more uncertain. For example, a sudden accidental release of large amounts of stored carbon dioxide could be catastrophic—possibly leading to a runaway greenhouse effect.

So to summarise, one plan could be that in the UK and other similar industrial countries, we run renewables mainly for electricity production, up to say 2020, and then switch some of the next wave of renewables over to hydrogen production for transport use after that. In addition, of course, some renewables will increasingly also be used for hydrogen production for the energy supply sector—for direct heating, and for electricity production in fuel cells. That all ties in well with the timescale for renewable energy expansion, since, by 2020, we should be reaching the point (a 20–30% electricity contribution from renewables) when it will be harder to integrate intermittent renewables into the national electric power system. The production of a storable gas solves that problem.

Of course this simple picture will be complicated by the likely development of new patterns of energy use—e.g. locally embedded generation, rather than central dispatch—but the general principles will still apply. The success of energy conservation and energy efficiency measures in each sector will also change the picture, although it seems unlikely that, even if we adopted all the technical fixes available for improving vehicle fuel efficiency, demand in the transport sector will be tamed. It seems bound to continue to rise—unless we impose draconian controls on private vehicle use.

The timing of the overall process will be shaped by economics of course, and by environmental legislation. If the environmental costs are imposed, for example by some sort of energy or carbon tax, by carbon trading, or by new Kyoto-style targets, then the whole process could speed up. A big unknown is whether Shell, BP and their ilk would be happy to give way to renewable hydrogen—certainly Shell seems to want to stick mainly with hydrogen from fossil fuels. But the sequestration option is going to be expensive, and become more so, as empty wells become scarce, whereas renewable hydrogen ought to be cheap by 2020 and get progressively cheaper. But it does all rely on us pushing renewables as hard as possible. There are some who say we can push them so hard that we could begin to take on the transport sector sooner rather than later. It is certainly true that, in principle, there are enough renewable sources out there to do it. But it will take time to get all the technologies up to speed.

You could say that this is just a technological approach, a way to sustain our addiction to ever-increasing mobility, and that what we really need are behavioural changes—less use of private vehicles, more walking, cycling, a shift to public transport. As usual the answer actually seems to be that we will need all of these changes, technical and social, if we are to tame transport demand. We will need changes in lifestyle and expectations concerning mobility, as well as in technology.

Air transport

That conclusion becomes even clearer if we look at air transport. Air traffic is responsible for around 10% of the total global warming effect (including around 3% of carbon emissions) and its impact is increasing. Technical improvements can certainly help increase fuel efficiency and that should limit emissions.

However, these gains can be undermined by changed patterns of use. For example, although the world aircraft fleet has doubled its fuel efficiency over the past 30 years, global air traffic has more than quadrupled since 1970, from 350 billion passenger miles to 1,500 billion passenger miles a year, and is forecast to more than double or even triple again by 2050. To put the issue starkly, while typically fuel efficiency has risen by around 3% p.a., in the year 2000 the use of aviation fuel in the UK rose by 10%. According to the Intergovernmental Panel on Climate Change, carbon emissions from this sector could increase by 478% between 1992 and 2050.[34] Unless demand is somehow curbed, it would require major technical fixes and efficiency improvements to reduce this.

Longer term, there have been proposals for a switch over to hydrogen-fuelled aircraft. Although it is lighter than conventional jet fuel, hydrogen has a low energy-per-volume ratio, so hydrogen fuelled planes would have to have large fuel tanks, and aircraft design would have to be radically changed. But if the hydrogen was produced from renewable sources, or in the interim from fossil fuels sources with carbon sequestration, that would obviously help reduce greenhouse gas emissions. Water vapour would be the only significant emission. However, since it would be injected directly into the troposphere or stratosphere (depending on the flying height), there could be problems, since water vapour plays a role in the greenhouse effect.

In the absence of any immediate technical fixes of sufficient scale to deal with the ever-growing emissions problem, there have been calls for aircraft fuel to be taxed. At present civil aviation fuel is one of the few commodities to have escaped taxation. Cheap flights have clearly been one reason for the boom in air traffic and consequent emissions. However, before adopting a completely anti-plane approach, it is worth noting that, perhaps surprisingly, modern aircraft, like the Boeing 777, consume only about as much energy per passenger mile on average as modern high-speed trains. So for some journey lengths there may be not much to choose between these options, at least in terms of fuel used and resultant emissions. However, there is still the point that, at present, aircraft fuel is not taxed, so that air passengers are not paying for their environmental impact.

In the final analysis, however, what matters in terms of energy use and consequent emissions is not the technical route adopted so much as the level of demand from consumers. Some technical fixes exist for some of the transport problems, but these may be undermined by consumer behaviour. That was what we have seen in the case of air traffic. It is also the case in terms of cars. The fuel efficiency improvements that have resulted from the relatively high level of fuel tax in Europe (compared with the USA) and the tight emission standards in the US, have been undermined to varying degrees by consumers' insistence on switching to inefficient recreational vehicles.

Clearly, resolving the transport dilemma is not going to be easy, given the extent to which existing patterns of transport are embedded in modern society. It has been said that the advent of the car has changed society more than most political changes; so if the car is now seen as a problem, then there is likely to be a need for significant social changes, as well as technical changes, in order to redeem the situation.

Chapter 4

The limits to sustainable energy

Energy efficiency and the rebound effect

The problem of dealing with rising demand for transport is just one example of the social problems that we need to face if we want to move towards an environmentally sustainable future. Technology can help, but on its own it is not sufficient. This is true even for what most people see as likely to be an unalloyed environmental good—energy conservation.

It is usually assumed that 'saving energy' is fundamentally a good thing environmentally, but also economically—it saves you money. Certainly there are many ways in which energy waste can be avoided, often at low cost. However there is a problem with this argument. To put it simply, the money that domestic consumers save by adopting energy conservation measures may be spent firstly on maintaining higher temperature levels for comfort, and then on other energy-intensive goods and services, like foreign holidays by plane. The result can be that at least some of the initial energy and emissions savings may be cancelled out.[35]

The exact scale of this 'rebound' effect is debated, with optimists suggesting that, since most products and services are less energy-intensive than direct energy use, the rebound effect associated with extra spending on goods and services might only account for a 10% reduction in emission savings that would otherwise have been achieved.

Nevertheless, overall it does seem likely that, all other things being equal, if any commodity becomes cheaper then more of it used. The optimists sometimes argue that there may be saturation levels for consumer demand in the affluent industrial countries, so that energy use may not continue to grow. Any new wealth will be spent on non- or low-energy consuming services, or on more energy-

---

Box 6: Energy Vampires

Many people are now used to the idea of buying energy-efficient lights, cookers, freezers, washing machines and so on, but they may be less focussed on the various electronic devices that now fill our homes—computers, videos, DVD players and the like. This could be a big, costly, mistake. Most domestic (and office) electronic equipment still consumes some power when it is ostensibly switched off, or in 'standby' mode. Some of this is because they have LED lights or timers which remain on, but most is due to the fact that many electronic devices actually run at low voltages and have a transformer to convert mains power (at 220-240 volts in Europe, 110 volts in the USA) to the necessary operating voltage (often 12 volts). The transformers consume a significant amount of power and, in some cases, remain in operation even when the device is not being used, as you can tell, since it gets warm. This is a very poor way to heat your home or office. Nationally, this hidden power drain can be the equivalent to the output of several large power stations.

In the USA these losses are known as 'Energy Vampires', and 'vampire slaying' has become a priority. This is not surprising given that it has been estimated (by Cornell University) that 'vampire' losses just from timers, clocks, memory and remote on and off switches, costs the US $3 billion p.a. and used the output of around 7 power plants. When all the other losses are added in, the scale of the energy waste is even larger. For example, in a speech to the Department of Energy in 2001, in which he evidently was the first to use the term, President Bush suggested that the overall vampire losses associated with computers, televisions and other common household and business appliances, cost the country 52 billion kilowatt hours of power annually, the equivalent of 26 power plants' output.

---

efficient consumer technology. However, so far, at the macro economic level, energy use by consumers in affluent countries shows no sign of decreasing, despite dramatic increases in energy efficiency. Moreover, as more people around the world join in the race to material affluence, demand seems certain to continue to increase.

This is not to suggest that energy conservation is not vital, since it is foolish to waste energy that has been produced at such a large potential environmental cost, and, in general, increased efficiency should make for economic competitiveness. There are certainly many opportunities for avoiding wasting energy—and money—in every sector of the economy. Energy savings of 50% and more are possible. Most people by now are familiar with the common sense idea of insulating houses well, but not so many will be aware of some of the other losses that they are paying for: **see Box 6**.

However, it does still seem inevitable that, even if we go all out to avoid wasting energy through clever technical fixes, and also manage to avoid the rebound effect, we will still come up against limits to what efficiency can achieve. For, as pressure for an improved quality of life from an expanding world population grows, the increased energy demand could outpace the increase in energy savings that can be made. The first wave of energy savings may be easy and cheap, but it is hard to see how sufficient increases in efficiency can continue to be made to offset the continuing rise in demand for energy. It may be that, at best, energy conservation, via the adoption of energy efficiency measures and demand management, will only be able to slow down the rate of increase in global carbon emissions. To actually reduce it requires a switch to non-fossil fuels—and even that may not be enough unless we also change our patterns of consumption and the expectations that underpin them. It is to that issue I now want to turn in the final section. If technology is not enough, what sort of social changes might we need?

A sustainable society

To recap the discussion so far, from some of the examples we have looked at, it seems that technology will not be sufficient to allow us to move to a sustainable energy future. Is that really the case? Surely there are plenty of renewable energy resources to allow us to escape the environmental constraints that limit the continued use of conventional energy sources.

As I have suggested, it is relatively easy to outline a series of 'technical fixes' for the climate change problem. Renewables can supply possibly 50% of world energy by 2050 and more later. Demand for energy can be dramatically reduced by clever Factor 4 energy efficiency gains, getting more use out of less energy, as promoted by Amory Lovins and co-authors in the seminal book

Factor Four.[36] But surely, even with clever technology, there must be some ultimate limits to growth on a finite planet?

Looking a long way ahead is obviously difficult. But some patterns are clear.

The chart produced by energy analyst Gustav Grob (see Fig 3) shows the relatively short period during which industrialisation occurred, based on fossil fuel. It's followed, after the projected demise of fossil fuels, by continued and accelerated expansion of energy use, based on renewables, up to about twice the current level of energy use. If true, that's good news. That period of expansion could allow the developing world space to catch up with the industrial countries, although of course, alternatively it could allow the industrial world to continue to expand ahead of the rest.

But either way, subsequently (according to this chart) growth can continue but not at such a rapid rate. Technical, ecological and land-use limits impose a natural limit on the amount of energy we can obtain from renewable sources—although we can raise this limit as we develop better renewable energy technologies and learn how to use natural energy flows more efficiently. Estimates vary as to what the limit actually is: some put it much lower than Grob, others much higher—maybe ten times more, or even 20. But the simple point is

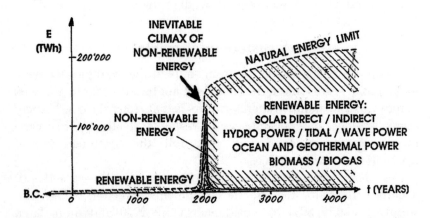

Figure 3: Grob's Chart
Source: Grob, G. (1994) 'Transition to the Sustainable Energy Age', European Directory of Renewable Energy Suppliers and Services, James and James, London.

that there are limits, and renewables can't help us escape these ecosystem limits.

Some people fear that this means that, sooner or later, we will have to face up to radical social, economic and cultural changes. Not everyone sees changes in lifestyles as a problem—some say that we would all benefit from a shift in emphasis from the quantity of consumption to the quality of consumption. And some say we should do this sooner rather than later, since the environmental and social problems associated with our current way of life are becoming urgent. Indeed, some say we have already gone beyond the viable ecological carrying capacity of this planet, and are living on borrowed time—borrowed from future generations.

But rising material expectations are locked into, reinforced by, and reinforce, the global system of economic expansion—we all seem to want more! Even some of the altruistically minded argue that global economic growth is the only hope for the developed world, if only in terms of allowing for some 'trickle down' to the less well off! With billions of new consumers potentially joining the race, as the developing countries industrialise, you can see why it's easier to think in terms of just changing the technology, and then just hoping for a more enlightened approach to consumption to emerge. That surely is not good enough.

To put it simply, it certainly looks as if environmentally sustainable technology can be developed and provide a technical fix for a while, but what we also need is to create a sustainable society—and that's a larger project. The big issue is: how do we go about it?

What future for the UK?

A series of four archetypal future scenarios has been produced by the Foresight team in the UK Government's Office of Science and Technology. They provide one way to discuss our options, and are summarised in **Box 7**.

According to the writers of the Foresight scenarios, the first, 'World Markets', is roughly how the world is going at present under the impact of globalisation, although some parts of it are adopting a more protectionist stance, as in the second scenario. By contrast the third scenario, 'Global Sustainability', is a world in which social and

---

Box 7: Foresight Scenarios

The first round of the Office of Science and Technology's Technology Foresight (TF) exercise, with 15 panels of experts looking twenty years ahead to see what technologies might be relevant, was completed, following a major consultation exercise, in 1995. In the energy sector, it was concluded that PV solar and nuclear decommissioning could be the main areas of growth.

A subsequent Foresight exercise, looking further ahead (forty years) culminated in a series of reports in 2000. In the new exercise, an Energy Task force was set up, which produced a consultative document, 'Fuelling the Future', which includes four socio-economic scenarios for the UK, to set the scene for a discussion of which technologies might be relevant.

The scenarios are, in much simplified terms:

1. World Markets—a world defined by private consumption and globalised high growth economies (what you might call the 'WTO' scenario for short!)
1. Provincial Enterprise—a protectionist, isolationist and nationalistic future with lower growth, interestingly, seen as being 'incapable of persisting for a significant period' (the 'Bush' scenario?)
3. Global Sustainability—ecological values enshrined in economic decisions and global environmental controls (the Kyoto scenario?)
4. Local Stewardship—a decentralised low-energy green community-centred future (Hobbit Socialism?).

It is relatively straightforward to see which energy technologies might fit well in to each of these (e.g. not much nuclear in 4), but there are some surprises. For example, it is suggested that there would not be any nuclear in 1, since it was not profitable enough, but there might be some in 2. Renewables figure in all of them, except possibly 2, with smaller-scale units being important in 4. The report suggests that, overall, one of the key technological requirements for all scenarios will be for more decentralised approaches to power networks, especially in 4. The buzzwords are familiar: dispersed, locally embedded micro power based on downsized renewables, hydrogen grids, and fuel cells/energy stores.

For details see http://www.foresight.gov.uk then look under Energy and Natural Environment publications. Or go direct via: www.foresight.gov.uk/servlet/Menu?id=4558&noredirect=y. You will find various versions of the Scenarios embedded in various reports. First off, there is the initial report on 'Environmental Futures' produced in

Box 7 continued   November 1998 for the Foresight project by the Science Policy Research Unit at the University of Sussex. The scenarios were then published by the Office of Science and Technology in March 1999 for use by the various Foresight panels, and, in November 2000, the Energy Futures Task Force set up by the Natural Environment and Energy Panel used them for their consultation document 'Fuelling the Future'. The Energy Futures Task Force then drew on the responses to this for its final report 'Energy for Tomorrow', published in August 2001. Meanwhile the PIU Energy Review team had started its work, and the scenarios were taken on board as a tool.

ecological values are more pronounced and in which the greater effectiveness of global institutions is manifested through stronger collective action in dealing with environmental problems. But it relies on something of a top-down approach—a managed technocratic, paternalistic future.

Finally, the fourth scenario, 'Local Stewardship', is a world where stronger local and regional governments allow social and ecological values to be demonstrated to a greater degree through the preservation of environments at the local level, with more emphasis on a bottom-up approach, with less focus on economic reductionism. In terms of energy, it's clear what this would imply: reliance in part on smaller-scale local renewable energy systems. But what sort of society would it be? What sort of values would it have—and how would they be induced and maintained? What would hold it all together?

Over the last few years there has been no shortage of analysis of the iniquities of globalisation and the need for an alternative approach to global—and local—economic development. Much of it stresses the need for 'localisation'—an economy in which decision-making power and wealth creation would be much more under local control, with a new emphasis on smaller scale, more self-reliant communities, rather than on international trade overseen by giant global corporations.[37]

Some elements of the 'bottom-up' localised approach have already been introduced in relation to regional planning for energy in the UK. Regional energy plans are being produced based on targets for the amount of renewable energy that can be produced.

The overall aim is to try to match the national target of a 10% contribution from renewables by 2010, but the Regional Development Agencies are being allowed autonomy to set their own targets, reflecting local conditions. So far, the whole thing is advisory rather than mandatory, but the local planning agencies will be under pressure to come up with realistic targets and then try to meet them by adopting appropriate planning policies, e.g. in relation to giving planning approval to wind farms and biomass projects.

It is not a big step from this to actually requiring agreed targets to be met with, say, access to central government funding for project development being made conditional on compliance. So they would be a combination of 'bottom-up' and 'top-down' approaches, with local autonomy and subsidiarity combining with central policing of agreed targets. So, it would be a mix of scenarios 3 and 4.

This approach can be used to support sustainable patterns of regional economic development. For example in Spain, acceptance of wind farm plans has been made conditional on the wind turbines being manufactured within the region. This has led to the growth of a distributed, rather than centralised, wind-power manufacturing industry.[38] Wind turbine manufacturing is well suited to this approach—wind turbines can be manufactured efficiently by medium-sized companies. Each region could probably meet a large part of its energy needs from its own energy resources, backed up perhaps by local hydrogen storage. But some areas would have much better renewable resources, and would be selling surplus energy to less well-endowed regions. There would also be trade between regions, for example, via the national power grid.

But there would be less emphasis on importing equipment—much of it would be manufactured locally. However, there would still be some specialisation, and an opportunity to export some of the products nationally and perhaps internationally. Equally, though, there might be a need for specialist imports.

Controlled trade?

This rough sketch of controlled trade in relation to energy systems and products might of course be seen as part of a wider approach to trade generally. However, the problems raised by trying to apply this approach to other commodities and services could obviously be very significant. Take food for example. Most people like regional variety, and imports from other areas are seen as desirable. It's the same for many other products—there are specialities and unique products that many people would prefer to have access to, but which may only be available as imports from other areas. In which case, how are we to avoid the creation of powerful and expanding companies, outside local control, who may be able to dominate local markets?

I do not want to get sucked into this debate too far—it moves us beyond the remit of this Briefing. But clearly there are implications for a sustainable approach to economics. The middle ground position is to impose tight controls over international trade, particularly in relation to environmental regulation. But, moving on to a radically revised (and you might say utopian) localised world economy, imports could be strictly limited to specialist goods and, similarly, exports would also be limited just to any excess volume produced—local markets would be met first. That would be one way to keep predatory external companies from gaining a foothold. Indeed, it might even be that import and export taxes are imposed, with the revenue raised used to fund a national agency, possibly linked with a global agency, whose job it is to resist the growth of global monopolies—a Sustainable Trade Organisation (STO) instead of a World Trade Organisation. It could provide funding to make local companies effective and carefully control any cross-boundary trade. There would be an emphasis on locally owned enterprises, possibly run as co-ops, with capital raised from local sources, aided when necessary by the STO, thus giving them an edge over external, predatory companies. So we would be adopting some of the 'economic protectionism' of Scenario 2.

This may seen a very draconian prescription—an end to 'free trade'. There will be the objection that it would fail, due to the removal of the benefits of open competition, which, allegedly, ensure that prices are kept low and more efficient technologies are devised. But a well-regulated and localised sustainable economy could have economic criteria which were just as strict, but different—more concerned with resource conservation and environmental protection. Optimal environmental productivity would be the

Box 8: Resource productivity and economic growth

As Amory Lovins and his colleagues have argued in the seminal book *Factor 4*, increasing the efficiency with which we use resources, including energy resources—making more with less—has obvious environmental and economic attractions. A report produced by the UK Cabinet Office's Performance and Innovation Unit in 2001 concluded that 'by improving resource productivity, we can cut costs and generate more value from finite stocks of non-renewable resource inputs'. It suggested 'by taking well-targeted action, implemented gradually and in consultation with those affected, the Government could jointly meet its economic and environmental objectives. This would in turn make a key contribution to delivering the goals of sustainable development.'

However, the final outcome will depend on precisely how you measure 'resource productivity'. The USA's response to climate change illustrates this point. In 2001, the US administration decided to withdraw from the Kyoto Climate Change accord, which called on the USA to reduce its greenhouse gas emissions by 7% compared with 1990 levels by 2008-2012. Instead it has adopted what has been called 'Koto Lite', an approach based on improving the USA's 'greenhouse gas intensity'. This is the ratio of the greenhouse gas emissions to the Gross National Product—i.e. tonnes of emissions per dollar of economic activity. President Bush announced that the aim was to reduce greenhouse gas intensity by 18% by 2012. He argued that 'This new approach is based on this common-sense idea: that economic growth is key to environmental progress, because it is growth that provides the resources for investment in clean technologies'. However, in effect, Kyoto Lite allows the USA to avoid making major changes in its energy system. The result could be an net increase in actual greenhouse gas emissions, over 1990 levels, of perhaps 14% by that time.

Obviously, reducing resource intensity in this way will reduce emissions from what they might have been, and some countries have actually done quite well in this regard. For example, whereas the US administration has often claimed that one of the reasons why they would not sign up to the Kyoto accord was because rapidly developing countries like China do not have to, in fact China has managed to reduce its greenhouse gas intensity much more rapidly than has the USA. But that still means that it is emitting more greenhouse gas year by year.

Moving to absolute reductions in emissions will require much more effort by all countries in the world. UK Prime Minister Tony Blair seemed

Box 8 continued to recognise the scale of the problem in a speech to CBI/GreenAlliance meeting in October 2000: 'The central theme of our approach is a more productive use of environmental resources. It is clear that if we are to continue to grow, and share the benefits of that growth; we must reduce the impact of growth on the environment. Some commentators estimate that we'll need a tenfold increase in the efficiency with which we use resources by 2050 only to stand still.'

So far, the UK has been quite successful in cutting greenhouse gas emissions without undermining growth. The UK's economy has grown by nearly 17% between 1997–2003 and, in that time, emissions have fallen by 5%, although, it has to be said, this has chiefly been the result of the switch from burning coal to burning gas for electricity production, which is cheaper. The next phase will be harder: demand for energy is still rising in some sectors and yet the UK has set itself the ambitious target of reducing carbon dioxide emissions by 20% by 2010, followed by the even more ambitious target of a 60% cut by 2050.

The PIU report 'Resource Productivity: Making More with Less', is available from the PIU website at:

http://www.cabinet-office.gov.uk/innovation, or go direct to
http://www.strategy.gov.uk/reports/reports.shtml

watchword, since that delivered a sustainable life style for the inhabitants of the area. That would still provide an incentive for Factor 4-type technical innovation—finding ways to get more from less, with less impact.

Is this sort of future completely utopian? Maybe not. Already we are seeing attempts to introduce resource and environmental productivity considerations into economic analysis. This was the starting point for the UK Cabinet Office study of energy futures, and it informs policies on many environmental issues. However, as Box 8 argues, the weakness of some versions of this resource productivity approach is that they assume continued emphasis on economic growth via ever increasing economic competition with other economic units—companies, nations. Control that, and it might be possible to avoid pressure for ever increasing exploitation of resources—a move nearer to a steady-state economy.

Of course, a weak point in this approach is that some consumers may still want the benefits of this ever-increasing exploitation of resources, regardless of its environmental impacts, and beyond the level attainable just by clever Factor 4 innovations. Education and normative pressures might limit rising expectations to some extent, but the only way to avoid pressures like this is for the system to deliver a lifestyle that is experienced as qualitatively better. More satisfying, more culturally and socially enriching, more equitable—and of course more sustainable. And that's a huge cultural and social challenge.

If the whole world is to 'contract and converge' to a sustainable level of consumption, albeit in a phased and equitable way, as Aubrey Meyer argues in his Schumacher Briefing (No. 5), then we have all got to learn how to live differently.

Chapter 5

What next?

The cost of energy

I have covered a lot of ground in this Briefing, ranging from relatively uncontentious technical issues to what some might seen as utopian fantasies. Utopias can have value in that they set out, or at least reveal, underlying goals and values. These can be debated and revised. But assuming there is a general agreement on the desirability of the overall concept of a switch to a sustainable energy future, what really matters is how we might head off in something like this direction.

Some of the necessary policy changes are needed urgently, and raise some key social and economic issues. For example, if we are seriously concerned about using renewables in order to help deal with climate change, we need to speed up the process of deployment. If for the moment some technologies are more expensive than conventional technologies, that means someone has to pay extra to help them develop. I want to conclude by looking at this cost issue, since it underpins many of the issues we will face as we try to move towards a sustainable future.

Higher prices for energy can be justified by environmental considerations. Why for example should clean renewables have to compete with dirty fuels like coal, or even gas? If we include the environmental damage costs of conventional energy sources in their prices, then renewables look much less costly.

The EU EXTERNE programme recently published its assessment of the prices that would emerge, if this process was carried out. The results are shown in Table 2, which shows the extra amount in euros that would have to be added to the current price of electricity in Europe to reflect the cost of the environmental damage caused by each energy source.

***Table 2:* Extra cost resulting from environmental damage in Euros/kWh**

(to be added to conventional electricity cost—assumed as 0.04 euro/kWh average across the EU)

Coal	0.057
Gas	0.016
Biomass	0.016
PV solar	0.006
Hydro	0.004
Nuclear	0.004
Wind	0.001

Source: EXTERNE (2001), 'Externalities of Energy' reports on the EXTERNE programme, European Commission, DG12, L-2920 Luxembourg, 1995, 1999, 2001.

As can be seen, the extra cost for wind is 4 times lower than nuclear, and 16 times lower than gas and 57 times lower than coal. Not everyone agrees with this analysis, or the data used, especially since it does not include the damage likely to be produced by climate change—since no one can predict how extensive that will be. But, on the basis of even these limited figures, at least the doubling of prices for electricity from coal-fired plants could be justified on environmental grounds, or, to put it another way, renewables could be condoned even though they cost twice the price of conventional power. The political issue is then: would consumers be willing to pay the extra?

The UK price obsession

A key policy issue is how to reflect these different environmental impacts in actual fuel prices. By taxes? That can be politically difficult. The UK government has found it hard to put extra taxes on fuels. It faced a consumer revolt in 1995 when it tried to impose full VAT on fuel and an even larger revolt in 2001 when vehicle fuel prices were seen as too high, with blockades of oil depots precipitating a major political crisis.

However, there may be a way out of this political impasse. Most people seem to object to taxation or price rises if they think the revenue raised is just going to disappear into the Treasury. However, if price increases are seen by consumers and voters to reflect specific environmental problems, and when the money is earmarked for specific projects in response, there has been less opposition. For example, the Non-Fossil Fuel Obligation, which, as we have seen, provided support for renewables via a levy on electricity, seemed to raise few objections. The technical term for this approach is 'hypothecation'—linking taxes to specific projects.

So far, this approach seems to be accepted by consumers in the UK, although the government has been cautious with the replacement for the NFFO, the Renewables Obligation, which was introduced in 2002. It requires energy supply companies to work towards obtaining 10% of their electricity from renewable sources by 2010, but, in effect, via a 3p/kWh 'buyout price', there is a ceiling on the price of the renewable energy that companies buy to meet the obligation. This is expected to limit electricity price rises to around 4% by 2010. The 2003 White Paper on Energy looked further ahead and suggested that the end result of the various measures that were being introduced to support the development of a sustainable approach to energy generation and use, might be an increase in electricity prices by between 5-15% by 2020.

Of course, making price predictions over such long periods is rather pointless—for one thing the base line prices of fossil fuel is likely to increase over this period. As it is, fuel prices in the UK are relatively low, in real (inflation adjusted) terms, about as low as they have ever been, and most people's electricity bills are usually very small—often smaller than their phone bills. People on a budget are obviously more seriously affected by any sort of price rise, but measures can be taken to protect the so called 'fuel poor'—in the extreme, cold weather payments, but more progressively by grants to help them improve the efficiency of their heating and energy use systems.

The fuel poverty issue is an important one and the UK government has begun to treat it seriously, as witness the coverage in the 2003 White Paper on Energy. Somewhat less creditably, the government has also bowed to pressure from industry, some sectors of which have complained about the cost of meeting another mechanism introduced by the UK government, the Climate Change Levy. This imposes a 0.43p/kWh surcharge on companies' use of power, but exempts those that use electricity derived from renewable sources. The environmental purpose of the Levy is clear, but some of the larger energy users, like Steel and Aluminium producers, have managed to get partial exemption from the Levy (in exchange for promising to reduce emissions), by arguing that they could not cope with the extra cost.

The UK seems to have a peculiar obsession with 'least cost' approaches—as if cheapness was the only relevant criteria. This has led to several problems outside the energy sector—for example, the cheap food policy led to shortcuts being made in food production and, arguably, to disasters like BSE (Mad Cow Disease). After that experience, most people now realise that food quality and associated farming practices are as important as the price of food. In the energy sector, the realisation that cheap energy may also be bad for your health and welfare is only coming slowly—as the impacts of climate change are beginning to be felt. For example, most people are still unwilling to face up to the pollution and emission problems created by their increasing use of cars and air travel, and, any attempt to raise fuel prices is resisted strongly.

That particular problem is not unique to the UK—it is even worse of course in the USA, where cheap vehicle fuel is seen almost as a birthright. But the emphasis on low prices has now also become a political maxim in the UK. It may have been the political success of the Thatcher government in convincing many people that competition automatically led to the best options emerging, with prices being seen as the key factor. This view stills seems to hold sway— and shows up in the energy field clearly. For example, in the UK there are a dozen or so green power tariff schemes, allowing consumers to support the generation or development of renewable energy, usually by paying a small surcharge on their bills. However the take up of these schemes has been very low—as noted earlier, so far only 60,000 consumers have signed up to get green power. By contrast the figures are 320,000 in Germany, 1.8 million in the Netherlands and 1 million in Denmark. Maybe the British are less environmentally concerned? [39] Actually, it's more the case that in the

UK the green power schemes were not advertised or pushed widely: the energy supply companies feared that, due to the slow development of renewables in the UK, there would not sufficient green power to meet demand. The situation was not improved by the fact that, under the rules of the Renewables Obligation, if energy supply companies want to run green power retail schemes with premium prices, they have to obtain green electricity that is additional to any electricity they claim against their Obligation. The result has been that the so-called 'voluntary' domestic green power market has been starved of green power. That's the price of the competitive approach adopted by the government, with the Obligation being seen as the main way ahead. By contrast, in some other parts of Europe, consumer take-up of green power has been one of the main drivers for the rapid development of renewables.

The UK government's obsession with low prices has also led to other major problems for renewables. In 2001 the government introduced New Electricity Trading Arrangements ('NETA') aiming to increase competitive pressures amongst energy suppliers—so as to force prices down. It has done that very successfully. Wholesale electricity prices fell by 20-25% in the first year of NETA's operation. However, the impact on renewable energy schemes has been disastrous—most are small companies who could not compete with the large conventional generation companies when prices were pushed down. The result has been that demand for their output has fallen by 44%. Part of the problem is that most renewables can only offer power intermittently and this is penalised heavily in the NETA structure—their environmental contribution is not recognised. So while the government says it is trying to push renewables forward, NETA is, in practice, pushing in the opposite direction.[40]

It was not just renewables that were undermined by NETA. Combined Heat and Power/co-generation projects were also badly affected. So too were coal plants and, mostly strikingly, the nuclear power industry. In September 2002, British Energy, the UK's main nuclear operator, had to ask government for financial assistance to avoid collapse. It was reported to be losing £4 for each megawatt hour sold. The government provided a £650m loan as an interim measure, but it seems clear that, whatever happens to British Energy, its nuclear plants cannot operate in the very competitive market

Table 3: EU Directive: 2010 Targets for Electricity from Renewables (%) excluding large hydro

Denmark	29.0
Finland	21.7
Portugal	21.5
Austria	21.1
Spain	17.5
Sweden	15.7
Greece	14.5
Italy	14.9
Netherlands	12.0
Ireland	11.7
Germany	10.3
UK	9.3
France	8.9
Belgium	5.8
Luxembourg	5.7
EU 15	12.5 %

created by NETA, without a significant public subsidy. If major, well-established energy suppliers like coal and nuclear cannot survive NETA, then new entrants like renewables may find it even harder to survive.

Supporting renewables

It could be that I have been too harsh on the UK. After all, it is trying to expand renewables, and to achieve the 10% by 2010/11 target will require nearly a 500% increase in capacity, involving the largest rate of expansion in the EU. That is mainly because the UK started from such a low level of renewable energy generating capacity, essentially a few hundred megawatts of hydro. By contrast Austria, Sweden and Finland enjoy very large hydro and traditional biomass contributions. As a result, in 1995 Austria obtained around 24% of its total primary energy from renewable resources, Sweden 25%, Finland 21%, while the figure for the UK was just 0.7%. The EU targets for electricity from renewables are shown in Table 3.

(Remember electricity is only a subset of total energy consumption.) As can be seen, despite its very large renewable resource potential, and the fact that this data leaves out large hydro, the UK still occupies a lowly position in this ranking. That seems to be the price paid for the emphasis on competition.

The UK's approach has been focussed on getting the prices for the leading renewables down, and as we have seen, it has succeeded in that, but at the expense of building up significant amount of generation capacity. By contrast much more generous subsidy arrangements exist some other EU countries, such as the REFIT 'renewable energy feed in tariff' system in Germany and elsewhere. The German REFIT system provides renewable energy generators with a guaranteed price, and the level of subsidy has been running at 8 cents or more/kWh for some projects. These arrangements have, unsurprisingly, led to the rapid deployment of renewable capacity.

The UK government's argument is that once the prices are right then expansion will begin. They may be proved right, but so far the REFIT fixed price type approach has clearly been more successful. Moreover, although REFIT approaches may have their problems, in terms of loading up utilities with extra costs, competitive pressures are not absent. Generators who can make use of cost-effective equipment will still be better placed than those that do not. Despite the REFIT type schemes in Germany and, until 2002, Denmark, there seems to have been no lack of technological innovation—quite the opposite as both have been at the forefront of wind technology innovation. The same can hardly be said of the UK.

The UK is not alone in using competitive market-based mechanisms to support renewables, and it is useful to compare how these schemes have fared compared with REFIT type schemes, with guaranteed price structures. A report by the European Environment Agency published in 2002 compared the successes and failures of EU renewable energy programmes between 1993 and 1999. It noted that three countries that guaranteed or fixed purchase prices of wind-generated electricity—Germany, Denmark and Spain—contributed 80% of new EU wind energy output during the period. This suggests that feed-in laws work better than the competitive tendering mechanism adopted by Ireland and the UK, a point reinforced by the problems subsequently experienced in Denmark when it switched from a feed-in tariff to a certificate trading system, which was later abandoned.[41]

A report by the World Wind Energy Association noted that more than 80% (1,144 MW) of the 1,388 MW of wind plant installed around the world in the first half of 2002 were installed in three countries with guaranteed minimum prices: Germany, Italy and Spain. In countries with quota/certificate systems, including the UK, USA, the Netherlands, and, more recently Denmark, only 75MW were installed. France and Brazil have decided to introduce minimum price systems which recognize the success of this framework. The Sierra Club, a US national environmental organisation, has also been campaigning for the adoption of feed-in tariff schemes in the USA.

The REFIT system has not been without its opponents—the European Commission would clearly prefer a market-led Europe-wide 'green certificate' trading system, and that may well come about in time. Certainly, REFIT type subsidies should not be needed to be retained across the board forever. They are useful at the early stage of a technology's commercial history, but they can be progressively withdrawn as it matures. Some have argued that tradable certificates can be just as effective in providing support, but so far, as we have seen, this has not been the reality. If and when a fully developed green certificate trading system emerges, following the planned (or rather sought for) harmonisation of the EU energy market, then perhaps matters might be different. But for the moment, a stepped/phased REFIT system seems like the better option.[42]

Conclusion

I do not want to suggest that competition and market-based assessments are bad things. They can ensure that resources are used efficiently, and, we certainly need to increase energy productivity for environmental as well as economic reasons. But obsession with low prices can be counterproductive when it comes to developing new energy technologies and developing a sustainable energy future. That is something the rest of the world can learn from the UK experience I have reviewed.

The problems with short-termism seem obvious enough. The UK

wave energy programme was halted in the 1980s because it was claimed that the price of electricity from this new technology would be very high—figures of 20p/kWh and even 50p/kWh were bandied about. Now we are having to try to catch up after having wasted twenty years. Solar photovoltaics has also been challenged on the same sort of grounds over the years. However it makes little sense to try to decide on longer term energy options using early estimates of costs based on prototypes or first generation technologies. Prices fall as the technology improves—assuming someone has the courage or foresight to fund its continued development.

As the UK Cabinet Office PIU report noted, there are well-established 'learning curves', showing how technologies improve in performance and cost over time. When performance is plotted against cost on a log-log scale (i.e. with the data plotted on logarithmic scales for both axes), a straight line results with a slope that varies from around 15-30%. The cost data in Table 1 earlier in this Briefing was based in part on using this form of analysis, with the learning curve slopes for PV and wind being put at between 18-20%, although for some of the newer technologies like wave power, more conventional engineering assessments had to be used.[43]

The simple message from learning curve analysis is that the rate of improvement can perhaps be improved with more money and effort. But one thing is clear: without sufficient support to move the technology on and to build a market for it, you do not get any improvement. That said, it does not appear that, despite very large-scale funding over a long period, nuclear power has successfully moved down a curve to low prices. Indeed the PIU reported that within the OECD, the nuclear learning curve slope was only 5.8%, which they attributed to the fact that it was a 'mature' technology involving large, inflexible projects with long lead times. They also argued that the frequent emergence of completely new designs meant that there was less technological continuity, less opportunity for economies of production scale, and less opportunity for learning. Either way, it does seem that nuclear power is an exception to the norm.[44]

To summarise, I have tried to show that there is a conflict between seeking low consumer prices and reducing environmental costs, at least in the short term. In the longer term, the new sustainable energy technologies should be able to expand to meet energy needs at reasonable costs, without imposing major environmental costs. The net cost to society could well be less, when we consider the huge potential costs of dealing with the impacts of unmitigated climate change. That implies that we should press ahead with developing renewables and other sustainable energy options, and not expect the new technologies to be instantly cheap. Some may well turn out to be—wind power looks to be one.

A devotee of free market competition would no doubt say that, in that case, all we have to do is leave it to the market to select the best options. But as I hope I have indicated, short term market assessments are not much use at looking ahead towards longer term social and environmental requirements, not least since they value the environment as a free resource. Instead we need to steer the market to ensure a sustainable future.

Market mechanisms, if sensibly steered, can help ensure that prices reflect longer term concerns. But of course that will only be part of the solution. We also need to win wide acceptance for increased prices for energy and, equally important, wide support for the new technologies. And beyond that, we will need to develop a new awareness of the need to move beyond a focus on the quantity of material consumption and on to a concern for the quality of consumption, in the context not only of environmental sustainability but also social equity. There are massive social and political implications to this vision which I have only begun to touch on in this short Briefing. What I hope I have managed to indicate is firstly that the energy technology is there to support a transition to a sustainable future, but secondly that on its own, technology will not be sufficient—there are some crucial social and political issues that will have to be addressed in the process of making the transition.

Epilogue: a solar world

No one can say how things will evolve in the years ahead, but some sense of what could emerge can be gleaned from a few final examples of programmes and projects around the world. Wind is the front runner. The major wind programme in Germany, aided by a progressive subsidy system, has already been mentioned, but other European countries are also pushing ahead—for example Spain already has 5GW of installed wind capacity and has overtaken the USA. India is now in third place, with over 1.5GW installed. Japan, a crowded series of islands with a relatively poor wind potential, nevertheless now has 500MW of wind plant in place. But solar PV power is beginning to catch up. The EU expects to have 3GW (peak) in place by 2010, the US is aiming for 3.2GW by 2020, while Japan, the world leader, is looking to have perhaps 80GW of PV by 2030. Meanwhile solar heating systems are already in widespread use around the world, and this area is still expanding rapidly. The European Solar Thermal Industry Federation has estimated that 1.4 billion square metres of solar thermal collectors could be installed in the EU—100 times more than the current capacity. New building regulations, specifying the use of solar devices, are a key driver for growth in the residential sector. Barcelona and several Spanish municipalities are implementing such regulations with great success. Biomass already provides 64% of the EU's renewable energy, mostly in the form of heat, and there are plans for dramatic expansion—energy crops and biofuels are seen as a key part of the EU's strategy. The development of efficient ways of using biomass in the developing world is also vital, and as noted earlier, there are some fascinating new ideas emerging, including the production of hydrogen gas from biomass via biological processes. But given the land use and planning problems of land-based energy extraction, perhaps the most interesting technological developments are in marine renewables—not just offshore wind but the new technologies for extracting energy from waves and from tidal currents. PV solar may well be the source with the largest long term energy potential, but the world also has very large offshore renewable resources.

Of course, in addition to new energy supply technologies and sources, we also need to push ahead with trying to slow the seemingly inexorable rise demand for energy, not least by developing more efficient ways of using energy. A lot of exciting initiatives are underway, many of them in response to regulatory changes. For example, in the EU and elsewhere, standards of performance are being set for a range of domestic energy-using devices, to which manufactures have to adhere, and local authorities are introducing energy considerations into local planning processes. But, however successful we are at energy conservation, whether through technology or lifestyle changes, it seems inevitable that, in order to respond to major environmental problems like climate change, we will have to deal with energy supply—and move to a world in which renewable energy dominates. We have to create a solar world.

Resources

In order that as many people as possible can influence the direction of the social and technological changes that lie ahead, there will be a need for much more information and education to be available on all aspects of the changes. Fortunately the internet provides us with a wonderful tool. There is a very large amount of material on sustainable energy easily available on the web—the problem is how the handle so much information, especially since the field is expanding rapidly. Just firing off a search via Google for some area that you are interested in could leave you drowning in an information overload.

There are some web sites which try to provide user-friendly ways of getting into the subject. Naturally I would recommend our own information service on renewable energy, *Renew On-Line*, which is a web version of parts of the journal *Renew* that we produce at the Open University: http://eeru.open.ac.uk/natta/rol.html.

In addition to a bi-monthly digest of news and commentary on technical and policy developments you will also find some introductory guides to sustainable energy on this web site. There is also an extensive guide to other relevant web sites.

If you are still wedded to print then there are two new OU text books which cover the field well: *Energy Systems and Sustainability*, ed. Boyle, Everett and Ramage, and *Renewable Energy*, ed. Boyle, both published by Oxford University Press in 2003.

For a more practical approach to renewable energy and sustainable lifestyles, see the excellent web site run by the Centre for Alternative Technology: http://www.cat.org.uk.

For links on local level sustainable energy developments around the world, see the Inforse network: http://www.inforse.org.

For a introduction to the climate change issue and some useful links to associated policy issues, as well as some practical steps you can take, see the Environmental Change Institutes web site at: http://www.changingclimate.org.

Some of the more general ideas discussed in this Briefing are explored in more detail in my book *Energy, Society, and Environment,* a second, enlarged edition of which has recently been published by Routledge.

References

1. The EU's Renewables Directive (2000), 'Green Electricity Directive Paper', May, and subsequent modifications.

2. Intergovernmental Panel on Climate Change (2001), 'Third Assessment: Climate Change 2001: Synthesis Report, Summary for Policymakers', Intergovernmental Panel on Climate Change, Geneva.

3. Jackson, T. (1992), 'Renewable Energy: Summary Paper for the Renewable series', *Energy Policy*, Vol.20 No.9, pp.861-883.

4. Elliott, D. (2003), *Energy, Society and Environment*, Routledge, second edition.

5. Department of Trade and Industry (2003), 'Our Energy Future: creating a low carbon economy', Energy White Paper, DTI London, Feb.

6. ILEX (2002), 'The Closure of British Energy's Nuclear Power Stations', report for Greenpeace, Nov.

7. FoE (2002), *Tackling Climate Change without Nuclear Power*, Friends of the Earth, London, Sept.

8. Royal Commission on Environmental Pollution (2000), 'Energy: the Changing Climate', 22nd Report of the Royal Commission on Environmental Pollution. The Stationery Office, London.

9. Ekins, P. (2001), 'The UK's Transition to a Low Carbon Economy', Policy Briefing PB5/01, Forum for the Future, London.

10. PIU (2003), 'The Energy Review', Performance and Innovation Unit, Cabinet Office, London.

11. As ref 5.

12. DEFRA (2001), 'Government looks for public consensus on managing radioactive waste', Press Release, Department for Food, Environment, and Rural Affairs, London, Sept 12.

13. Elliott, D. (2002), 'Land use and environmental productivity ', *Renew* 133, Sept/Oct, pp.22-24.

14. Elliott, D. with Taylor, D. (2000), 'Renewable Energy for Cities: Dreams and Realities', paper to the UK-ISES Conference on 'Building for Sustainable Development' May 26, RIBA, London.

15. Blowers, A. T. and Elliott, D. A. (2003), 'Power in the Land: conflicts over energy and the environment', in Bingham N., Blowers A. T. and Belshaw C.D., (eds) *Contested Environments* (U216 Environment Course, Book 3), Wiley/Open University Press.

16. Hewitt, C. (2001), 'Power to the People', Institute for Public Policy Research, London.

17. Roaf, S., Fuentes, M., and Thomas, S. (2001), *Ecohouse: A Design Guide*, Architectural Press.

18. European PV Industry Association and Greenpeace (2002), 'Solar Generation: Solar Electricity for over 1 Billion People and 2 Million Jobs by 2020'.

19. Laughton, M. (2002), 'Renewables and UK Grid Infrastructure', *Power in Europe*, 383, September.

20. Milborrow, D. (2001), 'Penalties for Intermittent Sources of Energy', Working paper for the PIU Energy Review, Cabinet Office, London. http://www.piu.gov.uk/2002/energy/report/working%20papers/Milborrow.pdf

21. Toke, D., 'Wind power needs CHP', *Renew* 140, Jan-Feb, p25.

22. As ref.10.

23. Grubb, M. (1991), 'The Integration of Renewable Electricity Sources', *Energy Policy*, Vol. 19, No. 7, pp.670-688, Sept.

24. Dunn, S. (2001), 'Hydrogen Futures: towards a sustainable energy system', WorldWatch Paper 157, Worldwatch Institute, Washington DC.

25. As ref. 5.

26. Shell (1995), 'The Evolution of the World's Energy System 1860-2060', Shell International, London.

27. See http://www.greenelectricity.org for updates on what is on offer in the UK.

28. World Commission on Dams (2001), 'Dams and Development: A New Framework for Decision-making', London.

29. Dougan, T. (2003), 'Marine Energy and Fishery Protection', *Renew* 143, May-June pp.25-26.

30. Karekezi, S. (2002), 'The case for de-emphasizing pv in renewable energy strategies in rural Africa', paper to the World Renewable Energy Congress, Cologne, June 29–July 5. Contact: African Energy Policy Research Network at www.afrepren.org.

31. Climate Care (2001), Carbon Storage Trust, quoted in *Green Futures*, July/Aug.

32. Shell (2001), 'Exploring the Future: Energy Needs, Choices and Possibilities: Scenarios to 2050', Shell International, London.

33. As ref. 2; and UN/World Energy Council 'World Energy Assessment: Energy and the challenge of sustainability', UN Development Programme, UN Department of Economic and Social Affairs and the World Energy Council, 2000.

34. As ref. 2.

35. Herring, H. (1999), 'Does Energy Efficiency Save Energy? The debate and its consequences', *Applied Energy* Vol. 63 pp.209-226. See also Herring (2000), 'Is Energy efficiency environmentally friendly?', *Energy and Environment* Vol.11 No.3 pp.313-325.

36. von Weizsacker, E., Lovins, A., Lovins, H. (1994), *Factor Four*, Earthscan, London.

37. Hines, C. (2000), *Localisation: a global manifesto*, Earthscan, London; Douthwaite, R. (1996), *Short Circuit*, Green Books, Dartington.

38. Connor, P. (2001), 'Wind Power in Spain', EERU Report 077, Open University, Milton Keynes.

39. See http://www.greenprices.com.

40. EAC (2002), 'A Sustainable Energy Strategy? Renewables and the PIU review', House of Commons Environment Audit Committee, Session 2001-2002 Fifth Report, London, July 2002.

41. EEA (2001), 'Renewable energies: success stories', Environmental issue report No 27 : Ecotec Research and Consulting Ltd and A. Mourelatou, for the European Environment Agency, Copenhagen 2001.
http://reports.eea.eu.int/environmental_issue_report_2001_27/en

42. EIGreen (2001), 'Action Plan for a Green European Electricity Market', EIGreen project report (WP7); Claus Huber, Reinhard Haas, Thomas Faber, Gustav Resch, John Green, John Twidell, Walter Ruijgrok, Thomas Erge; Energy Economics Group (EEG), Vienna University of Technology, Austria.

43. PIU (2002), 'Generating Technologies: Potentials and cost reductions to 2020', Working paper, Performance and Innovation Unit, Cabinet Office, London.
http://www.piu.gov.uk/2002/energy/report/working%20papers/PIUh.pdf

44. PIU (2002), 'The Economics of Nuclear Power', Working paper, Performance and Innovation Unit, Cabinet Office, London.
http://www.piu.gov.uk/2002/energy/report/working%20papers/PIUi.pdf

OTHER SCHUMACHER BRIEFINGS

The Schumacher Briefings are carefully researched, clearly written booklets on key aspects of sustainable development, published approximately twice a year. They offer readers:

- background information and an overview of the issue concerned
- an understanding of the state of play in the UK and elsewhere
- best practice examples of relevance for the issue under discussion
- an overview of policy implications and implementation.

No 1: Transforming Economic Life
A Millennial Challenge
James Robertson

Chapters include Transforming the System; A Common Pattern; Sharing the Value of Common Resources; Money and Finance; and The Global Economy. Published with the New Economics Foundation. ISBN 1 870098 72 2 **£5.00 pb**

No 2: Creating Sustainable Cities
Herbert Girardet

"An excellent briefing"
—Scientific & Medical Network

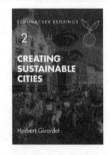

Shows how cities can dramatically reduce their consumption of resources and energy, and at the same time greatly improve the quality of life of their citizens. Chapters include Urban Sustainability, Cities and their Ecological Footprint, The Metabolism of Cities, Prospects for Urban Farming, Smart Cities and Urban Best Practice. ISBN 1 870098 77 3 **£5.00 pb**

OTHER SCHUMACHER BRIEFINGS

No 3: The Ecology of Health
Robin Stott

Concerned with how environmental conditions affect the state of our health; how through new processes of participation we can regain control of what affects our health, and the kinds of policies that are needed to ensure good health for ourselves and our families. ISBN 1 870098 80 3 **£5.00 pb**

No 4: The Ecology of Money
Richard Douthwaite

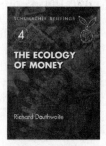

"This slim volume crams in a wide range of ideas about money" —Sustainable Economics

Explains why money has different effects according to its origins and purposes. Was it created to make profits for a commercial bank, or issued by government as a form of taxation? Or was it created by users themselves purely to facilitate their trade? This Briefing shows that it will be impossible to build a just and sustainable world until money creation is democratized. ISBN 1 870098 81 1 **£5.00 pb**

No 5: Contraction & Convergence
The Global Solution to Climate Change
Aubrey Meyer

"Hidden within this short book is a proposal which could and should alter the course of history." —The Ecologist

"A simple yet powerful concept that may yet break the deadlock of climate negotiations." —Red Pepper

The C&C framework, which has been pioneered and advocated by the Global Commons Institute at the United Nations over the past decade, is based on the thesis of 'Equity and Survival'. It seeks to ensure future prosperity and choice by applying the global rationale of precaution, equity and efficiency in that order. ISBN 1 870098 94 3 **£5.00 pb**

OTHER SCHUMACHER BRIEFINGS

No 6: Sustainable Education
Revisioning Learning & Change
Stephen Sterling

"A stimulating, challenging and thought-provoking book."—School Science Review

"A valuable contribution to the literature on education and sustainability."
—Environmental Education Research

Education is largely behind—rather than ahead of—other fields in developing new thinking and practice in response to the challenge of sustainability. The fundamental tasks are to • critique the prevailing educational and learning paradigm, which has become increasingly mechanistic and managerial • develop an ecologically informed education paradigm based on humanistic and sustainability values, systems thinking and the implications of complexity theory. An outline is given of a transformed education that can lead to transformative learning. ISBN 1 870098 99 4 **£5.00 pb**

No 7: The Roots of Health
Romy Fraser and Sandra Hill

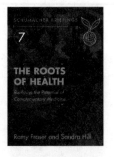

The advancements of modern medicine provide a sophisticated but mechanistic approach to health. Dazzled by its progress, we have lost touch with the simple remedies and body wisdom that were once a part of every household. By understanding the roots of health, we can begin to reclaim this wisdom, and to heal ourselves, our society and our environment. ISBN 1 903998 05 0 **£5.00 pb**

Schumacher Briefings are available from The Schumacher Society (details on following page). All bookshop orders and foreign rights enquiries should be sent to Green Books (details on back cover).

OTHER SCHUMACHER BRIEFINGS

No. 8: BioRegional Solutions
For Living on One Planet
Pooran Desai and Sue Riddlestone

"Some people talk about sustainable development — others just get out there and do it! BioRegional does just that, coming up with inspiring but practical initiatives that address climate change, waste reduction and the all-important challenge of regenerating local communities and economies sustainably." —Jonathon Porritt, Programme Director, Forum for the Future, Chairman, UK Sustainable Development Commission

In this Briefing Pooran Desai and Sue Riddlestone show how we can meet more of our needs for wood products, paper, textiles, food and housing from local renewable and waste resources. They outline the theoretical framework of bioregional development and the award-winning practical solutions that BioRegional have developed with industry partners. They quantify how we can significantly reduce CO_2 emissions and recycle waste, and so greatly reduce our ecological footprint. ISBN 1 903998 07 7 **£6.00 pb**

No. 9: Gaian Democracies
Redefining Globalisation and People-Power
Roy Madron and John Jopling

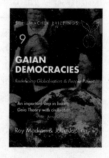

The Global Monetocracy which has evolved over the past three hundred years is engaged in the relentless pursuit of economic growth in order to sustain a global debt-money system, for the benefit of the wealthy and powerful. This Briefing shows how Gaian Democracies could achieve very different aims. Throughout the Briefing, the authors stress the systems framework on which they have based their proposals, and draw on examples to show how many of the principles of Gaian democracies have been successfully applied in the real world. 160pp ISBN 1 903998 28 X **£8.00 pb**

THE SCHUMACHER SOCIETY
Promoting Human-Scale Sustainable Development

The Society was founded in 1978 after the death of economist and philosopher E.F. Schumacher, author of seminal books such as *Small is Beautiful, Good Work* and *A Guide for the Perplexed*. He sought to explain that the gigantism of modern economic and technological systems diminishes the well-being of individuals and communities, and the health of nature. His work has significantly influenced the thinking of our time.

The aims of the Schumacher Society are to:

- Help assure that ecological issues are approached, and solutions devised, as if people matter, emphasising appropriate scale in human affairs;

- Emphasise that humanity can't do things in isolation. Long term thinking and action, and connectedness to other life forms, are crucial;

- Stress holistic values, and the importance of a profound understanding of the subtle human qualities that transcend our material existence.

At the heart of the Society's work are the Schumacher Lectures, held in Bristol every year since 1978, and now also in other major cities in the UK. Our distinguished speakers, from all over the world, have included Amory Lovins, Herman Daly, Jonathon Porritt, James Lovelock, Wangari Maathai, Matthew Fox, Ivan Illich, Fritjof Capra, Arne Naess, Maneka Gandhi, James Robertson, Vandana Shiva and Zac Goldsmith.

Tangible expressions of our efforts over the last 25 years are: the Schumacher Lectures; the Schumacher Briefings; Green Books publishing house; Schumacher College at Dartington, and the Small School at Hartland, Devon. The Society, a non-profit making company, is based in Bristol and London. We receive charitable donations through the Environmental Research Association in Hartland, Devon.

Schumacher UK Members receive Schumacher Briefings, Schumacher Newsletters, discounts on tickets to Schumacher Lectures & Events and a range of discounts from other organisations within the Schumacher Circle, including Schumacher College, Resurgence Magazine and the Centre for Alternative Technology (CAT).

The Schumacher Society, CREATE Environment Centre,
Smeaton Road, Bristol BS1 6XN Tel/Fax: 0117 903 1081
admin@schumacher.org.uk www.schumacher.org.uk